1,001 WAYS TO SAVE THE PLANET offers something to aim for. No one can do everything at once, but this book will give you ideas. Do things you feel comfortable with at first. Take it one step at a time. If we all did just a small percentage of the things listed in this book, we would go a long way toward solving some of our most pressing problems. Each of the 1,001 things described are short and simple. They may save you money. They may encourage you to stop doing something you always suspected was wrong or wasteful. They may change your life-style. And they will definitely help you become a conscious, aware person, ready to save the planet!

BERNADETTE VALLELY

1,001 Ways
to Save
the Planet

Bernadette Vallely

IVY BOOKS • NEW YORK

Ivy Books
Published by Ballantine Books
Copyright © 1990 by Bernadette Vallely
The moral right of the author has been asserted

All rights reserved under International and Pan-American Copyright Conventions. Published in the United States by Ballantine Books, a division of Random House, Inc., New York. Originally published in Great Britain by the Penguin Group in 1990.

ISBN 0-8041-0757-2

This edition published by arrangement with Viking Penguin, London

Manufactured in the United States of America

First Ballantine Books Edition: September 1990

To Susan

Contents

Introduction

It's heartening to see so many people ready to do positive things in favor of the planet.

Around the world today we want information to help us make the right environmental choices. Over 50 percent of us are willing to buy environmentally friendly products, and we are beginning to question the use of cars, not just because of poisonous emissions but because of the very cars themselves. A smaller percentage of us are willing and able to pay higher prices for organic foods, and as their prices fall, more of us will buy them to stop the overuse of pesticides. We are willing to buy unbleached, recycled paper to save trees and to stop the pollution of our river systems. We have forgone aerosols for something less damaging to the environment.

We are beginning to realize that the damage to the environment is related to *us*—the way we live and the things that we consume. It's not happening "out there" in space. It's happening right here, in our homes and workplaces. Our reaction is crucial in the fight to save the planet, and we can't ignore it any more.

1,001 Ways to Save the Planet is deliberately "dark green"; it offers something to aim for. No one can do everything at once, but this book will give you ideas. Do things you feel comfortable with at first. Take it one step at a time. If we all did just a small percentage of the things listed in this book, we would go a long way toward solving some of our most pressing ecological problems.

Each of the 1,001 things to do listed here are short, and most of them are simple. They may stimulate you to get in touch with a group or organization or to read other books, but in the main they represent straightforward actions that most people can take.

They may save you money.

They may encourage you to stop doing something you always suspected was wrong or wasteful.

They may change your life-style.

They may all add up to your becoming a conscious, aware person, ready to SAVE THE PLANET!

Acknowledgments

I am indebted to Charo Lamas, who has helped me type and retype the manuscripts with joy and inspiring competence, and to Sandra van der Feen for spending time on the telephone to check facts and figures on my behalf.

Thanks must also be given to those who have supported me with love and enthusiasm and who have read and commented on the manuscript, suggesting their own ways to save the planet, supplying me with statistics and references, information and ideas, and educating me on many environmental issues that I knew little about: Sue Adams, Stewart Boyle, Dave Buffin of the Pesticides Trust, Miranda Castro of the Spectrum Clinic, Pat Fleming, Alison Costello, Guy Dauncey, Liz Cook of Friends of the Earth US, Ann Link, Jan McHarry, Brigitte Mierau, Jose Pepper, Debbie Silver, Hugh Singh of Plantlife, Tessa Strickland of Penguin Books and my mother, Susan Vallely.

Acknowledgments are due to the hundreds of organizations and individuals who have helped me to prepare this book, from those who supplied literature and information sheets to those who helped me dig deep to find answers that might not otherwise have surfaced. Readers of this book should be aware of the dedication, professionalism, and sheer stamina that make up just some of the qualities of those who work in the environmental world. They include: Animal Aid; Stuart Bell, Pension Investment Research Consultants; Michael Birkin; Liz Bisset, Islington Borough Council; Mary Blake, Friends of the Earth UK; Anne Brant, Balloons & Clowns Ltd, USA; Hartwin Busch; John Button; Giles Chitty, Barchester Green Investors;

Karen Christensen; the Civic Trust; Madeline Cobbing, Greenpeace UK; Sue Coffin, National Magazine Company; the Conservation Foundation; Keith Corbett, the Herpetological Conservation Trust; the Council for the Protection of Rural England; Simon Counsell, Friends of the Earth UK; Adrian Davies, Friends of the Earth UK; Nigel Dudley; Energy Efficiency Office; Environment Defense Fund, USA; James Erlichman; Kharis Fausset, Friends of the Earth UK; Ferguson's Building, N. Ireland; Herbert Girardet; David Goldbeck; Pad Green, Friends of the Earth UK; the Greenhouse Crisis Foundation, USA; the *Green Magazine*; the Haldane Trust; Tony Hare, Plantlife; Jonathon Hopkins; Liz Howlett; Johnson Wax; Briony Jones; Julia Langer, Friends of the Earth Canada; the London Food Commission; Debra Lynn-Dadd; LYNX; Robin Maynard, Friends of the Earth UK; Kai Millaird, Friends of the Earth Canada; Fiona Murie, London Hazards Centre; Natural Resources Defense Council, USA; David Newick, Greencard BCC Card Centre; Andy Ottoman, Greenpeace UK; Pauline Pears, the Henry Doubleday Research Association; Gord Perks, Greenpeace Canada; Colin Redman, National Deaf Children's Society; James Richardson, Anti-Apartheid Movement; Royal Society for Nature Conservation; Royal Society for the Prevention of Cruelty to Animals; Carole Rubin, Canadian Environmental Network; Anne Scott, Royal Society for the Protection of Birds; Seventh Generation Trading, USA; Catriona Slovach, Pension Investment Research Consultants; Michelle Sweet, Pollution Probe Canada; Alistair Townley, GreenGauge; United Nations Environment Programme; the Vegan Society; the Vegetarian Society; Joan Walley, MP; Clive Williams, Faculty of Environmental Urban Spaces, Polytechnic of North London; World Wide Fund for Nature, Jeffrey Hollender, seventh Generation Trading; Joyce Auaub, Skin Cancer Foundation; Mothers Against Drunk Driving; Steve Kaufman, The World Watch Institute; People for the Ethical Treatment of Animals; Humane Society; Fund for Animals; Heidi Prescott, National Audubon Society.

Particular thanks go to the organization Common Ground, whose work I find inspiring, poetic, and unique and to whom

I owe much of the reaffirmation of my own love of the countryside and, especially, the beauty and nobility of trees.

Half of the proceeds from this book will go to the Women's Environmental Network, an independent, nonprofit organization committed to educating, informing, and empowering women who care about the environment. Women and men are equally welcome to join, and for more information, or if you want to share your own ideas on how to save the planet, please write to the Women's Environmental Network, 287 City Road, London, EC1V ILA. Please include a stamped addressed envelope or $5 if you need a reply.

THE GREEN HOME

The Kitchen

Use a "green" kettle

A kettle is an important part of your kitchen, and you're likely to use it several times each day. To save energy, buy an automatic kettle that can boil as little as one cup, or use a small kettle on your stove.

Descale your kettle

Descaling your kettle prevents the buildup of fur inside and around the element, which in turn leads to a longer boiling time and wasted energy. You will find that filtered water stops the kettle furring up so quickly. Use descalers that are nontoxic and not tested on animals. These are now relatively easy to find. Try using a strong vinegar solution as a cheap and simple alternative. Rinse the kettle out carefully afterward.

Use small pans for cooking

Use the smallest pan possible for cooking food. Don't waste energy on heating up the metal of an oversized saucepan. Turn down the heat to match the size of the pan.

Use steamers and dividers

You don't need three saucepans to cook three different types of vegetable. One good steamer or pan divider will do. You then need only one ring or burner, and steamed vegetables taste far better than boiled ones and are more nutritious.

Use saucepan lids

Put the lids on your saucepans. There are two advantages: lids save energy, as water and food boils more quickly; they also reduce the condensation in your kitchen, which leads to dampness. Turn the heat down once water has boiled, saving further energy. Cook with the right amount of water. Most foods need only enough water to cover them.

Stop running the water

Using running water to wash your dishes is extremely costly and wasteful. It is far better to get a bowl or to use a sink plug.

Use cold water

Use cold water for rinsing dishes, washing vegetables, and completing a host of other kitchen tasks. This will save the precious energy it takes to heat the water and will keep your fuel bills low. You could put a note near the taps to remind you.

Ventilate your kitchen

If you are using a gas stove instead of electric, you will be saving lots of money on your energy bills. But you could be storing up trouble. According to American scientists, young children especially could be at risk from respiratory illnesses resulting from badly ventilated kitchens in which gas is used for cooking. Open the window or use a solar-powered ventilating system.

Clean your tumble dryer

The tumble dryer uses large amounts of energy to keep a drum circulating hot air around with clothes in order to dry them. A small lint screen filters the air that enters the drum. Clean it regularly. A screen with any fluff or wool on it will make the machine work harder, wasting energy.

Service your fridge

Every household has a refrigerator, but most people don't realize that these machines gobble up energy, using valuable resources and contributing to acid rain and global warming. Servicing your fridge can make a major difference to your electricity bills and to the performance of the machine. Check the seal for leaks or loose rubber gaskets; clean the coils at the back of the machine at least once a year; use a thermometer to check the temperature inside the refrigerator; and look out for tell-tale signs like a frost building up too quickly or the walls of the unit "sweating." Defrost the machine several times a year, and never let frost build up to more than ¼ inch (5 millimeters). Cover all foods, especially liquids, to keep frost down. Proper attention to your fridge will keep it efficient, save energy, and extend its working life.

Design your life to save energy in the fridge

Don't put your refrigerator near the stove or other hot appliances. Don't leave the door open longer than necessary and buy only as much food as will fill the shelves with ease, so that air can circulate. Remember to turn the temperature down to save energy in winter. Get advice about the correct temperature. By thinking a little in advance, you could save money on your bills and save energy at the same time. Design yourself an energy-efficient life-style.

Dispose of your old fridge carefully

Old fridges can be a hazard. CFCs can leak, and that is when they cause the most damage. These substances are very stable and will last until they reach the ozone layer. Then they break down, releasing chlorine. Ask your local authority or the manufacturer to recycle the CFCs safely. Throwing your fridge out could also be a death trap for small children, who could get caught inside and suffocate. Insist that the machine is disposed of with consideration to the environment.

Buy a CFC-reduced, energy-efficient fridge and freezer

If you are about to go out and purchase a new fridge or freezer, there are two major considerations.

First, it must be energy efficient. In the United States refrigeration accounts for 7 percent of the total electricity used and one third of the average household energy bill. Check for new energy-saving features. Most companies now produce at least one model. The Consumers' Association gives details of running costs. The difference in your electric bill between the best and worst models can be up to $50 for fridges and $170 a year for freezers. For the environment it can mean a lot more.

Secondly, your fridge or freezer must not threaten the ozone layer. Buy a model that is equipped with less CFC insulating foam.

Check your halon fire extinguisher

Forty percent of all household accidents happen in the kitchen, and a great deal of those are fires. Five experts in the United States suggest that 16 percent of fatal fires begin in the kitchen, and at least 300,000 such fires are thought to occur every year. Although a fire extinguisher is an important part of your safety equipment, halon fire extinguishers are known to be serious ozone depletors. It is obvious that we have to be alert to the dangers of fires in the home, but for electric, grease, and liquid fires (like hot fats) buy a halon-free extinguisher or a good fire blanket that will deprive a fire of oxygen. Do learn how to treat fires responsibly, and teach other members of your household, too.

Change your boiler

Over half the fuel bill in an average home is generated by heating the house and the water. A well-maintained gas central-heating system uses less energy and produces less pollution than either an electric or a solid-fuel system. The most efficient type of boiler available is the gas condensing boiler, which extracts heat from the waste flue gases because it allows them to condense in the boiler. These boilers cost more, but if your current model is old, or if you are moving to a new home, the investment could save you 10 to 15 percent of your fuel bills. Modern controls and regular annual maintenance of your boiler could save you as much as 35 percent of the bills.

Get rid of cockroaches safely

Cockroaches are nasty insects to find at home or in your workplace. They make holes in piles of packaging or near pipes and breed very quickly. Traditionally they have been sprayed with pesticides, but there are more acceptable and surer methods of getting rid of these disease-carrying insects. (Their eggs are protected in capsules, so they can often survive pesticide spraying.) Gaps of more than ¼ inch (5 millimeters) need to be cemented, but the most obvious expedients are to keep your house clean and to check under and behind surfaces, especially behind fridges. To kill cockroaches use a mixture of flour, plaster of paris, powdered sugar, and borax or bicarbonate of soda, which cockroaches will eat. They also seem to be attracted to sweets, so glass jars containing sweets could trap them.

Recycle bottles

We use 28 billion glass bottles and jars every year. Those that are recycled include beer bottles, which are recycled many times, and milk bottles, which are used an average of twenty-four times before being crushed and melted into glass again. By recycling you save oil—each ton of waste glass uses 30 gallons (136 liters) less oil than new glass—and therefore precious resources. Use bottle banks.

Use recycled glass

Light-green glass for drinking glasses, jugs, and kitchenware gives you the perfect recycled product. It is slightly thicker than normal glass and is tough and hardwearing.

Insulate your dishwasher

There are no standards that regulate the insulation of dishwashers, which means they can be inefficient, especially smaller models without even basic coverings. This is especially true in cold climates or when a dishwasher is installed against an outside wall, as the temperature outside may affect the water temperature. If you can't buy a model with thick insulation, then consider boxing it in yourself. The tiny cost will be quickly recouped through greater efficiency, and you will also have a quieter machine.

Use your dishwasher the greenest way

It is arguable that using dishwashers can never be green. They use less water—about 4¼ gallons (20 liters) a day versus the 9–14 gallons used when washing dishes by hand for a family of four—and less energy (1.5 kwh versus an average of 2–3 kwh), but when the energy needed to manufacture the machines is taken into account, there is little saving. What is clear, however, is that dishwashers save time and labor. You can make sure that your dishwasher is as green as possible by making sure you always have a full load, selecting a temperature at least 50 °F (10 °C) lower than recommended, washing large dishes by hand, and using a cold-fill option to save energy.

Plan small meals carefully

If you live on your own or often have to cook for yourself, you may find that a multicooker or slow cooker, which uses as much electricity as a light bulb, could save you money and energy. You may choose to cook a large amount and freeze or store some for another meal later on. This will also save energy.

Don't use a microwave

There is a widely held belief that a microwave can save you energy, but a recent study by the Canadian government shows that you save a grand total of $10 a year in electricity costs if you use a microwave rather than an electric stove. Only 35–40 percent of the energy microwaves use is actually used to cook the food; the rest is wasted. A recent study in a British medical journal, the *Lancet,* found that microwave cooking of milk, cheese, meat, and fish creates potentially dangerous abnormal amino acids that are neurotoxic, which means that they may kill brain and nerve cells and are toxic to the liver and kidney. A British government report in 1989 found that one-third of the ovens tested failed to heat food uniformly to a temperature that will kill bacteria. As microwaves are intended primarily for preprocessed foods, the health-conscious greenie will do well to avoid them.

Beware of the pretty dish

Glazed pottery dishes, fruit bowls, and food containers may contain lead and cadmium. Concern has been raised in Europe about the continued use of toxic metals in glazes used by potters who make their own pots and in those that come from developing countries. Once again there is a double problem— the lead and cadmium are harmful to the environment but also potentially harmful to us through our food. Buying industrial pottery from sources you can trust is usually a good bet because the glazes will have been sealed.

Open the windows

Instead of using artificially created air fresheners from aerosol cans, try a greener alternative. Use potpourri mixes of herbs and dried flower petals for a natural smell. Most air fresheners in aerosol cans use chemicals like limonene and imidazoline, which work by blocking the sense of smell and giving off a pungent odor that masks everything else. They are suspected carcinogens in animals. Aerosol companies may have removed the CFCs, but they use hydrocarbons that are ''soft'' green-house gases instead. Hydrocarbons also contribute to local smog. Aerosols are not biodegradable and take more energy to produce than they are worth. If you cannot easily find pot-pourri, try opening a window!

Use food containers

Instead of wrapping everything in cling wrap, aluminum foil, and paper, get some good sturdy food containers and save on waste. Save energy in your fridge by covering food with a lid.

Don't throw your waste down the sink

Sink waste-disposal units are energy intensive and encourage bad housekeeping. Save your food scraps for the compost heap instead of adding them to our polluted waste systems.

Washing up your dishes

Use the smallest possible bowl in which your dishes will fit. Save time and energy by steeping them in water first to loosen difficult stains.

Cadmium-free plastics

Cadmium is an acutely toxic heavy metal, and there is evidence that even tiny quantities have reproductive effects. Once taken out of the ground, concentrated, and released into the environment, it stays there forever. The millions of kitchen bowls, storage bins, dustbins, brushes, and pans that we buy each year are made from plastics that may contain cadmium; bright-red and orange plastics are the main candidates. Because of pressure from the European community the production of cadmium will be curtailed in coming years. Some companies have already removed cadmium from their plastics, and you should look for the cadmium-free label before you buy.

Store and reuse plastic carrier bags

The first lesson is to learn to say NO to carrier bags, but if you have them, do not throw them away. Reuse them as rubbish bags.

Do not use paper towels

We buy billions of kitchen towels in the United States, where even dishcloths and tea towels have been replaced by the paper throwaway. The paper comes from trees and is bleached using chlorine, unless the labels say otherwise, which means the pollution, in the form of dioxins and furans, of our waterways. Use clean cloths to wipe up spills, and rinse them frequently. Do not be fooled by advertising that tells you that a cloth is unhygienic.

Get good lighting

If you have a big kitchen, you are likely to spend quite a lot of time in it, and you will want to fit the best lights for the environment. Energy-efficient lighting produces up to 75 percent less carbon dioxide, the main greenhouse gas, than ordinary bulbs. Good lights for the kitchen will be more expensive initially, but within a year you will be saving money.

Look after your kitchen

Kitchen utensils should be kept clean and in good condition. Cheap or badly made utensils just won't last and will cost you more in the long run. Catering shops are a good place to find well-made utensils.

Find ecological soap liquid

Washing liquids are generally made from petrochemical-based derivatives. Waterways around the world have been polluted and destroyed by oil slicks and secondary pollution from the petrochemical industry. Avoid fragrant and colored washing liquids; they have usually been tested on animals. Campaign for your favorite brand to label its ingredients clearly. Use brands whose labels declare them to be cruelty-free and to contain ingredients that you can trust.

Use a pressure cooker

Save energy when cooking: use a pressure cooker for vegetables, stews and tasty, quick meals.

Make jams and chutneys

A friend of mine is trying to campaign for a two-week public holiday in early September so that everyone can help pick fresh fruit and vegetables and begin pickling, making jams and chutneys, and preserving in syrup. Mrs. Beaton wrote in 1859, "To be acquainted with the periods when things are in season is one of the most essential pieces of knowledge which enter into the art of cookery." Her recipes include apple jam, damson cheese, and quince jelly. Fruits in season are certainly much cheaper, and by making our own preserves we can save money, use up the jam jars we have been collecting and produce a really wonderful alternative to sugar-laden jams from the supermarket.

Get rid of useless gadgets

Some gadgets that are foisted upon us at birthdays, Christmas, and weddings use up energy and waste resources. Do not buy them if you have a choice, and think carefully about their usefulness. Surprisingly, the coffee maker and electric frying pan come out badly in terms of energy use, and the waste compactor, waffle iron, and electric toaster are not far behind. An electric blender uses very little energy and costs a few cents to operate, so it could make a worthwhile utensil. Forget electric carving knives and specialist gadgets unless you are cooking for a restaurant.

Compost your kitchen waste

Collect the organic matter that you throw away each week in a bucket or tub and compost it in your garden or patio for a rich fertilizer. Feed scraps to pets, and save some for the birds, too.

Switch to gas

Natural gas is over 40 percent more energy efficient than electric heating or cooking. It makes economic sense to switch to a better fuel.

Wrap up without aluminum

Aluminum foil comes from minced bauxite, some of which is derived from destroyed rain-forest areas. It is extremely energy intensive to produce, using nearly 6 tons of oil for every ton of aluminum. Reuse tubs and pots for storing food, and use dish covers made from glass for the oven.

Check your pots and pans

Aluminum was first used for cooking by the French nobility of the early 1800s, but as the price fell, it soon lost its glamor and everyone was able to cook with the wonderful new metal. An increase in aluminum in the body has been linked to diseases such as Alzheimer's disease. Many people feel that as the aluminum in pots and pans has a chemical reaction, especially when acidic foods like tomatoes and rhubarb are cooked, we should avoid them and opt for stainless steel, cast iron and enameled cookware. Don't use aluminum teapots either, as tea itself is said to contain high levels of aluminum.

Pack your freezer

You can save energy and help your freezer to run better if it's filled to capacity. Buy foods that are cheaper in bulk, and make several dishes together in the oven to save energy and money.

Green up your kitchen units

Chipboard is still used for many of the world's kitchen units these days. The wood is chipped, then stuck back together in long sheets with formaldehyde. About one in five people are thought to be sensitive to formaldehyde, and the gas seeps from

chipboard products for many years. Do not buy tropical hardwood either. Go for a locally produced hardwood if you can
afford it, second-hand units or soft wood (such as pine) from a
sustainable source.

Use recycled black plastic bags

Every year we spend billions of dollars on black plastic trash
bags to collect the rubbish and waste that we have accumu-
lated. Ninety-six percent of our rubbish is technically recyclable, and if we recycled as much as we could, there would be
less need for black bags. Until that happens, we will continue
to use them, so at least try to buy trash bags made from recycled plastic.

Separate your waste

All over the United States, Canada, and Britain, we are waking
up to the fact that we consume and waste precious resources.
The average household throws away over 2 tons of rubbish
each year, creating a mountain of stinking, rusting, chemical-
filled garbage. A large proportion of it could be recycled.
About 30 percent of our rubbish is paper and card, easily
recycled; another 10 percent is glass, with excellent recycling
potential; another 30 percent is organic, which can be com-
posted. The rest is plastics, metals, and packaging made from
two or more types of material, for which recycling is difficult,
if not impossible. On average 3 percent of the waste is babies'
disposable diapers. By learning to separate your waste in the
kitchen you will start to create the climate for successful recycling programs, and you can already do a lot without formal
programs. The main point is to be conscious of the waste that
you put into the trash can and to think of the effect that its short
life has had on the environment.

Rain forest–free kitchenware

An acre of tropical rain forest is destroyed every second. The
forests represent the richest source of life on earth, maintaining
half of all living creatures. One of the ways in which we

destroy the rain forest is by demanding millions of kitchen bowls, knife handles, bread boxes, and cutting boards each year. If you buy these items, always check where the wood has come from. A hardwood is more likely than not to have come from a forest, so avoid the product if you are not sure. Small companies are now emerging that produce goods made from the wood of sustainably forested trees. These are more environmentally friendly.

Disinfect without harm

When I grew up I used to think that disinfectant came from hospitals and was a type of medicinal cleaner. I never thought that it could be poisonous or toxic, and yet today's disinfectants contain a number of volatile chemicals like trichlorophenol, cresol, benzalkonium chloride and formaldehyde. Far better to use borax and hot water and simply to keep your household as clean as possible. One American hospital monitored bacteria for a year while using borax and found that it completely satisfied all its germicidal requirements and cost less, too.

Use vinegar to clean glass and mirrors

Save yourself money: use vinegar, water, and paper for a sharp cleaner that is better value than the heavily marketed aerosol sprays that release tiny drops of irritating ammonia. You may have to rub hard at first to get rid of the buildup of the old cleaner, so be prepared for a second clean.

Always use the minimum

Whether you are washing or cleaning clothes, surfaces or dishes, a good environmental rule is to use the absolute minimum to get the job done. We often apply cleaners so liberally that most of the product goes down the sink or drain. Save energy, avoid pollution, and always used the minimum.

Use lemon juice

Lemon juice can be used as a mild bleach, for cleaning metals and copper and for descaling. You do not have to buy bottles of lemon juice; the lemon itself comes in its own biodegradable packaging! Save the rinds to put around your roses to ward off cats.

Clean your oven without having to use aerosols

Use a paste of baking soda and water for a nontoxic alternative to sodium hydroxide- and hydrocarbon-filled aerosols when you clean your oven. A container topped with a pump-action spray and filled with soap (green, of course) and borax in warm water is a good alternative cleaner.

Try your own dishwasher soap

If you are dissatisfied with the dishwasher powders that you are offered, particularly because they contain strong chemicals and are mostly tested on animals, try making your own. For a fairly good alternative, use borax from a pharmacy or hardware store mixed with half as much baking soda.

Get rid of tea stains with baking soda

Use the cheapest of all scouring powders, baking soda, to get rid of tea and coffee stains in cups and on surfaces. Ninety-five percent of the average powder cleaner is chalk, with added perfumes, bleaches, and detergent. Chlorinated powdered bleaches produce fumes when wet and can be irritating to the eyes, nose, and throat. As well as baking soda, you can use borax or even table salt as a good abrasive.

Use a water filter

Clean, safe water should be on tap for everyone, but persistent use of nitrates and pesticides, combined with neglect by the water industry, has resulted in higher and higher levels of bacteria, heavy metals, lead, and aluminum in tap water. Vital for the lives of billions of people and animals across the world, we have to campaign for it not to be polluted, overused and wasted. Buy a water filter if you can afford it, but everyone should demand cleaner water.

Grow spider plants

The spider plant, or *Chlorophytum*, has been a favorite house-plant for over two hundred years. It is fast growing and virtually pest-free. Even those who don't know anything about plant care can usually manage to grow a spider plant, which needs lots of water in spring and summer and very little in winter. Recently, however, it has been discovered that *Chlorophytum* can absorb formaldehyde vapors, making it a "pollution absorber." You might consider growing spider plants if you have allergies and think you are susceptible to formaldehyde.

Return all returnable bottles

Some of the bottles you buy are returnable. An efficient system operates in bars for the collection of beer bottles, for example. But you can also return some medicine bottles to pharmacies and soft-drink bottles to local shops. Don't leave empty bottles to gather dust; why not spend an hour returning them on your next day off?

Buy hard-wearing rubber gloves

Don't waste resources by buying throwaway gloves. Thicker and more hard-wearing ones will last far longer, although you may have to look in a hardware shop for them. Good gardening gloves are essential, too, and should last for years.

Use cloth napkins

Paper napkins are a wasteful and expensive use of trees and energy. Cloth napkins have been around for centuries: choose them as a washable and reusable alternative. Paper napkins are generally bleached with chlorine, adding to pollution in the place where they are made, and the use of trees for such a disposable product is questionable. If you travel, get into the habit of taking your own napkin with you.

Save aluminum

Save ring pulls from cans, milk bottle tops, and clean aluminum foil for collection by local charities that sell it for recycling.

Use old tea for your plants

Feed your plants with leftover tea from the pot. I've fed mine with tea for years, and they love it. (Don't use milk or coffee, though.) You can also put tea leaves on the compost heap.

Keep your kitchen clean

One way to ensure that you are not visited by mice, rats, and infestations of insects is by keeping your kitchen clean. Cleaning means looking under the refrigerator, cleaning behind the stove and picking up things when you drop them, as well as making sure that you sweep and wash surfaces and floors regularly. I use baking soda on the floors and surfaces and sometimes a little rosemary essential oil for a fresh, sweet smell.

The Bathroom

Cultivate spiders

There are hundreds of thousands of types of spiders, and they all seem to end up in your bathtub. But spiders are an important part of the ecosystem. They live and eat all the creepy crawlies you don't like—houseflies, earwigs, cockroaches, and other insects that can spread disease. They are, in turn, an important source of food for birds. Most of them are completely harmless and very small. The one in your bathtub is usually the male *Tegenaria gigantea*, looking for a mate. He has fallen into the tub from the ceiling and been trapped. Turning on the water will drown him. Cultivate spiders: don't kill them.

Avoid chemical toilet cleaners

Toilets, as we know them, were invented in Scotland ten thousand years ago. The inhabitants of the Orkney Islands built the first latrine system, a dug-out channel system designed to carry waste from the home to a nearby stream. They were aware of the dangers of their own waste even then. Today we put bleaches, scourers, cleaners, sprays, and toilet blocks into our toilets to keep them clean. Natural microorganisms exist that break down and dispose of human excrement. By using so many chemicals we actually delay that process. Toilet blocks that go inside the cistern can contain paradichlorobenzene, which is highly toxic if swallowed and is reported to have caused cancers in laboratory animals. It is persistent in the environment. Use vinegar-based toilet cleaners that have not been tested on animals.

Turn off the tap

Dripping taps waste enormous amounts of energy. A fast-dripping hot tap could fill a bath tub every day. Taps should be repaired if they are faulty, and rubber washers should always be in good condition.

Take a shower

The average bathtub holds 30 gallons (136 liters) of water. The average five-minute shower uses only 15 gallons (68 liters). Over a year, millions of people taking showers instead of baths could save billions of gallons of hot water, which in turn would save energy and reduce pollution and sewage problems.

Share a bath

If you cannot use a shower in your home, you can at least share your bathwater, which would cut the amount of energy needed to heat and produce the bathwater by half! This would be as energy efficient as taking a shower.

Use pure soap

Most dermatologists agree that pure soap and warm water is the best way of cleaning the skin. Fragrances, tonics, and other added ingredients can irritate the skin and can even cause allergies; they also, of course, add to the stockpile of synthetic chemicals and manufacturing processes that are polluting our environment. Strong substances remove the skin's layer of protective oils along with the grease and grime.

Scrub your skin with a washcloth

Use a cotton washcloth to remove grit and bacteria without chemicals. You may find it a better alternative to the expensive sea sponges that come from the Mediterranean and strip the sea of its natural marine life.

Make your own talcum powder

Talcum powder bought in a shop may well be contaminated with cancer-causing asbestos fibers. This is because the talc is often mined in the same geological areas as asbestos. You can make your own talc with cornstarch, arrowroot or oat flour, and add ground-up herbs or dried flowers for natural fragrances.

Don't throw your sanitary wear down the toilet

Sanitary pads and tampons are labeled "fully flushable," and the manufacturers maintain that they can be easily flushed away. But throwing them down the toilet is not the last of pads or tampons. Most of our sewage is untreated and so goes out to sea without any treatment at all. That means that literally billions of sanitary pads and tampons are left to float around in the water. They take about 120 days to biodegrade, and the plastic backing strips do not biodegrade at all. They are washed up on beaches and get caught up in fishing nets, and, as they contain bacteria and fecal contamination, they make a pretty bad meal for an inquiring seabird. Wrap yours in paper bags manufactured for the purpose and put them in the trash can.

Don't buy rain forest bathroom fittings

After pine, the most popular bathroom fittings come from tropical rainforests. Mahogany, a rich, luxurious wood, is used to make toilet seats, shelving, towel racks, mirror frames, and toilet-roll holders. Major department stores around the world promote these fittings, contributing directly to rain forest destruction. When a mahogany tree is felled, up to a hundred others may be cut and burned simply because they are not as valuable financially. Don't buy rain forest fittings. Stick to alternatives like pine or glass.

Go for squeaky-green hair

Most shampoos are made from the same ingredients, mainly synthetic detergents, as soap. If you wash your hair, then rinse it, you are likely to wash off the herbs, perfumes, oils, and added extras. Hair does not generally need conditioning creams either. Do not buy shampoos that have been tested on animals.

Mix up your own hair conditioner

If you need hair conditioner, instead of using synthetically produced hair conditioners why not mix up your own? Plain yogurt mixed with egg and a quarter teaspoon of nutmeg makes a really wonderful conditioner. No plastic, no chemicals, no additives.

Beware of dandruff shampoos

The antidandruff market is big business. To avoid the embarrassment of dandruff, we allow our scalps to be poisoned with chemicals like selenium sulfide, cresol, formaldehyde, and resorcinol. Most of these are toxic substances that can easily be absorbed through the skin, can burn eyelids, and can even cause drowsiness and unconsciousness. They are tested on animals and produced in huge quantities only to be poured down our sinks into the water system. Professor David George from the University of Utah says, "Dandruff is not a disease. Most people have it."

Brush your teeth

Recent studies have shown that high nitrate levels found in food and water used as fertilizers to grow vegetables and fruits can be dangerous to your health. In the body nitrates can form cancer-producing nitrosamines, and the World Health Organization has recommended that we reduce our intake. One report suggests that up to 60 percent of the nitrates absorbed by our bodies is produced by food bacteria in the mouth. Regular brushing and flossing of teeth can reduce the intake.

Clean your teeth with cleaner paste

Most toothpastes on the market today contain all kinds of artificial chemicals to keep the dentist away, but they may not be all that good for us. Sugar substitutes such as saccharin are known to cause cancer in animals, yet they are added to toothpaste to give it a sugary taste. Other ingredients may include ethanol, formaldehyde, ammonium, titanium dioxide and artificial flavorings and colorings. Titanium dioxide is used as a whitener in toothpaste. It has led to acidic pollution in river systems, and Greenpeace has been campaigning to end its production completely. Many natural toothpastes, without any of these ingredients, are available from health-food stores and supermarkets.

Turn off the water as you brush your teeth

You don't need to let the water run while you're brushing your teeth. It's just a habit. Get into the greener habit of turning off the tap as you are brushing. If the whole nation did this, we could save many millions of gallons of water.

Forget the electric toothbrush!

We buy 350 million toothbrushes in the United States every year, but only a small percentage of us go as far as buying an electric toothbrush. It does not use a lot of energy, but the batteries are wasteful, needing fifty times as much energy to produce as they provide. They don't really clean teeth any better than ordinary toothbrushes and are a waste of money and resources.

Nontoxic shaving cream

Don't use aerosol spray cans of shaving cream if you can help it. They are filled with chemicals that are ozone depleters or greenhouse gases and are made from synthetic chemicals. Try a cheaper nontoxic alternative with light vegetable oil or plain soap.

Cut out the disposable razor

Believe it or not, according to archaeologists early man shaved himself with sharpened flints and shells as long ago as 20,000 years B.C. The throwaway variety didn't appear until much later, in 1903, when King Gillette, a failed door-to-door salesman, hit upon the idea when he was shaving one morning. Every day in the United States, 5 million disposable razors are produced, making them a profitable but wasteful use of energy and raw materials. Use an electric shaver, or grow a beard and cut out the disposable razor.

Keep a tub of baking soda in the bathroom

Don't bother to use scouring powders or creams in the bathroom. Your bath, tiles, and sink can be cleaned with baking soda or borax with no extra effort.

Insulate your hot-water tank

Although most of our homes have insulated hot-water tanks, much of the insulation is thin and badly fitted. You will find that a thick insulation jacket for your tank will reduce your electricity bills and save energy; the payback is usually within weeks.

The Living Room

Use extra-thick curtains

If you sew a backing on your curtains to make them extra thick, you'll find that they will prevent the heat in your living room and bedroom from escaping through the window. Curtain lining is much more cost-effective than double glazing, and floor-to-ceiling curtains keep even the coldest temperatures at bay.

Block up your fireplace

According to US statistics, 8 percent of the heating of our homes escapes through the chimney. If you're not using your fireplace, it should be blocked up carefully, so long as you don't block all ventilation, which would cause condensation.

Turn off the remote control on your TV

Millions of remote-control television sets are thought to be switched off at night by the remote control button alone, leaving the TV to use a quarter of its power all night long. This unnecessary and lazy habit helps to generate thousands of extra tons of carbon dioxide each year, according to Friends of the Earth, which wastes energy and money.

Watch green TV

Television sets create millions of tons of carbon dioxide, a greenhouse gas, yet we could be using more energy-efficient televisions. If you have to buy a television, ask about the relative electricity consumption. Don't buy those with remote controls.

Recycle magazines

Every year in the United States we buy over 20 billion magazines, made from millions of trees. Magazines printed on glossy paper can be difficult to recycle at present, but you could be creative by taking them to your local doctors' office, hospital, or dentist.

Use cornstarch or baking soda to clean your rugs and carpets

Powdered carpet cleaners are usually baking soda with added perfume, packaged in an attractive way. Liquid shampoos are potentially more dangerous, as they contain such chemicals as perchloroethylene, a known human carcinogen, and naphthalene. These are extremely toxic and should be used only as a last resort. Try shaking ordinary baking soda or cornstarch over your carpets or cleaning them with a mixture of white vinegar and boiled water for stubborn stains. Ordinary dish detergent is good as a shampoo.

Polish your furniture with olive oil

For wooden furniture that has been badly stained, try polishing with olive oil instead of using aerosols with phenols (possible human carcinogens) and nitrobenzenes. Mix the oil with white vinegar and water for varnished woods. Save the ozone layer, reduce global warming and energy consumption, and save yourself some money, too.

Avoid foam sofas

Polyurethane foam ignites easily and gives off a thick black smoke that contains cyanides. Literally thousands of people have been killed or injured in foam fires, and fire fighters and safety officials advise against their use without added protection. The foam itself is known to cause cancer in animals.

Recycle your telephone

Without a doubt, the telephone is our most important mode of communication. Telephones can be purchased in shops on almost every street. Every year British Telecom recycles about 4.5 million telephones. They are stripped, and the electronic components are reused. Metals such as gold, silver, and palladium are recycled separately, and the 1,700 tons of granulated plastics are melted for other uses. Instead of throwing away your telephone, send it back to the manufacturer for recycling.

Go without the phone book

Telephone directories are made from precious paper, yet many of us never need them. Computerized operator systems can give you most numbers cheaply and effectively, and you may find that all you need is the *Yellow Pages* or your local community or business directory. This could cut down on the number of phone books that are given free to households every year. Apparently the glue used on the spines of telephone directories make them harder to recycle. You may like to take that up with the telephone company.

Go for negative ions

Negative ions are really positive. They can make all the difference to the atmosphere in a room. Many people use air ionizers to improve the amount of useful negative ions in the room, especially if they suffer from asthma, hayfever, bronchitis, or migraines. A negative voltage refreshes a room and wakes everyone up. You can achieve the same effect by opening the window, burning a candle, or vaporizing essential oils.

Moisten the air

Dry living rooms with central heating can block your nose and create a stuffy atmosphere. Place small bowls of water or *commercial humidifiers* under the radiators *to act as humidifiers*.

Buy secondhand furniture

Buying secondhand furniture is an environmentally sound form of shopping. You can often pick up great bargains—pieces that you wouldn't have been able to afford new. Visit auction centers (they aren't as frightening as they at first seem), house auctions and sales and secondhand or nearly new shops. Let your friends know that you wouldn't be averse to some extra furniture, and they'll often be only too happy to oblige. Don't be frightened of putting on a handle or fixing a hinge. Simple jobs can take you minutes and often save you money.

Lighting up the planet

Lighting in the living room is important. You can't afford to strain your eyes when you are reading, for example, but you can buy special lamps and clip-on lights that use compact fluorescent bulbs. Although they cost more initially, they will save you money within a year and cut carbon dioxide levels that cause global warming. Just eight ordinary light bulbs of 60 watts produce one ton of carbon dioxide during their lifetime. It takes forty-six fluorescent bulbs to create the same amount of pollution.

Make your flowers last longer

Everyone wants cut flowers to last longer and look beautiful, and you don't have to use chemicals on them. I find that cutting the stems into an angle with a very sharp knife helps the flowers take more water. Be prepared to change the water and keep the vase full. Apparently the fungicidal properties of a brass pot or a penny keep the plants fresh longer, although I haven't tried this myself.

Site your thermostat in your living room

The central-heating thermostat controls the whole system and the temperature throughout the house. You could be making your system work overtime if it is positioned near windows or external doors that cool it down. The end result will be overheating in many rooms. Better to place the thermostat away from drafts in the room you use most often, such as the living room.

The Bedroom

Have a rain forest–free bedroom

Many bedrooms use wood for wardrobes, beds, and drawers. Less expensive chipboard furniture is bonded with formaldehyde resins and may have hardwood veneers obtained from rain forests or plastic surfaces that attract dust and dirt. Solid-wood bed sources are easier to identify, and the beds will last longer, as they are sturdier.

Buy a sound bed

Your bed can be environmentally friendly. Buy a bed from sustainable forest sources. That means not buying hardwoods, but softwoods such as pine are fine. You can even find organic, animal-free waxes and varnishes to finish the bed and give you a green night's sleep.

Don't buy cheap foam mattresses

Foam mattresses can be fire hazards, releasing toxic cyanide fumes when burnt, but they also can contain CFCs, responsible for destroying the ozone layer. Go for natural fillings for your mattress instead. Concern has been voiced in West Germany and around Europe about magnetized spring coils in mattresses. Doing without them may help us to have a better night's sleep, as our nerves will not constantly have to readjust to the magnetic impulses from the springs.

Use a hot water bottle

There's something very comforting about a hot-water bottle. It can warm more than your feet, and I've known it to help relieve period pains, cramps, and the misery of flu. Use it instead of an electric blanket. Some evidence has emerged of an increase in the number of miscarriages among women who use electric blankets, and cancer is suspected.

Check your bedding

Rich, thick duvets and fleecy pillows can make your bed comfortable and warm, but do you know where they come from? Synthetic materials are made from petrochemicals, a nonrenewable, nonbiodegradable resource; some duvet fillings are made with traditional duck feathers, which are biodegradable but are increasingly taken from intensively reared ducks, kept in appalling conditions. You will have to make up your own mind before you can sleep easy in your bed.

Keep warm with natural fibers

Blankets with acrylic fibers are produced from acrylonitrile, which is a suspected human carcinogen. The chemical has been known to cause breathing difficulties and weakness in those sensitive to it. It might be better for your health to avoid such blankets and opt for those made from nontoxic alternatives such as pure wools and cottons.

Rest your head on a toxic-free pillow

Polyurethane foams used in pillows can be dangerous to your health. The foams are produced from petrochemical derivatives, and studies in the United States have found higher instances of bronchitis and skin problems among people who are in contact with them. Polyurethane foams are known carcinogens in animals, but the main concern centers on their behavior in a fire. When mixed with polyester (often used in pillowcases), they can produce toluene disocyanate, a poisonous gas. Most deaths from fires occur because of smoke inhalation, not burning, and having such foams in your home will increase your risk of serious injury if there is a fire.

Buying an iron-free solution?

Some sheets and linens are sold on the basis of being "easy-care," "no iron" or "permanent press." This usually means that formaldehyde has been added. The manufacturing process ensures that the formaldehyde is permanent and will release its fumes continuously. It can aggravate asthma and cause coughing and streaming eyes. Mixtures of polyesters and cottons often have formaldehyde finishes. If you have suffered at night, it may be because of your sheets, so take a closer look at what you lie on.

The Hall

Use compact fluorescent lighting

The electric light was invented between 1878 and 1879 by a Briton and an American. Joseph Swan patented his light in 1878 in England and Thomas Edison his in 1879. They had both discovered how to use carbon to make a small, heated filament glow. These first bulbs lasted only some 150 hours; now ordinary incandescents can last about 2,000 hours. Compact fluorescent lights are the same size as an ordinary light bulb but can last for at least 8,000 hours and only use one fifth of the energy for the same amount of light. Although they are more expensive initially, compact fluorescents can also save you money.

Spring-green your light bulbs

Lights attract dust and dirt, which in turn make your lighting less efficient, so at least twice a year dust and clean light bulbs all over the house.

Use outdoor lights rarely

Outdoor lights should be used only where they are really necessary, as in the case of potentially dangerous steps, a dark road, or an unusually positioned front door. Spotlighting for large houses is particularly wasteful. You should always use long-lasting, compact fluorescents. You could link them with a timer set to go on as dusk falls and to switch off when the sun comes up.

Insulate your letter box

Gaps and drafts can easily occur around your letter box. You should think about insulating it with a brush or flap.

Use a letter opener

The average American receives 598 pieces of mail every year (most of that is what we call junk mail). Even if the mail you receive is not junk, you can make use of the envelopes and paper. Use a letter opener and collect the empty envelopes in one place. I recycle the large ones and use small ones as notepaper for messages. Stamps can be collected and sent to help charities.

Reject junk mailings

When you join an organization, or you fill in a questionnaire in your local store, you could be unwittingly allowing yourself to be besieged by junk mail because your address may end up on a mailing list that is sold for profit. In America 100 million trees are cut down every year to be used for junk mail. When you join new organizations, ask them not to sell your address. Write to the mailing preference service to get your name removed from its lists. In the case of companies that persist in sending you their junk, I suggest sending it back to them without a stamp: they may start taking note of your requests more quickly!

Put a note on your front door to stop free newspapers

Millions of free local newspapers are dropped through our doors every week. You do not have to be on a mailing list to receive this unsolicited garbage; people are paid to deliver them straight to your door without using the postal system. My local free paper has the audacity to look like a newspaper on the front page, with news stories and information, but open it up and every single page is filled with advertisements. Put a note on your front door: say NO to wasted paper. If they still arrive, start writing letters, not just to them but to the real local press, too.

Check your front door

Every year we buy millions of doors made from wood grown in tropical rain forests. For the front door you will need a hardwood, but it is best to buy from native sources. Ask builders, do-it-yourself stores and builders' merchants where the wood comes from, and avoid chemical treatments if you can.

Buying from representatives

The "Avon lady" has become synonymous with door-to-door selling in America and Britain. Avon has secured a large proportion of the makeup and beauty market with exclusive ranges not normally available in the shops. American research has shown that Avon reps earn less than the federal minimum wage. They are, basically, a free labor force for the company. Other products often sold on doorsteps or in especially arranged parties are cleaning products, household goods, shower units and home improvements such as double glazing. Do not be taken in by the hype. Ask yourself if you really need the product offered. Ask representatives to call back in a week's time before you sign anything, so you can "cool off" and decide without pressure.

Buy your doormat from a cooperative

Supporting cooperatives from developing countries is one way of ensuring that handicrafts, art, and local industry are kept alive without exploitation. You can buy wall hangings, doormats, prints, and handicrafts and know you are supporting

businesses that care about workers and share the wealth among
them.

Recycle your newspapers

If you recycled your own copy of *The New York Times* (or a
similar newspaper) you could save 4 trees, 15 pounds of air
pollutants from being pumped into the atmosphere, 2,200 gallons of water and 1.25 million BTUs of energy. Contact your
local recycling center. Save your newspapers for recycling,
and if you have nowhere to send them, start applying pressure.
Why not share your newspaper with a neighbor or read it in the
library for free? Don't buy those you don't really read, and
request a recycling service to be provided by your local council. Buy more recycled paper products to support the market.

Do-It-Yourself

Assess the energy you use in your home

It is vital that you assess the energy that you use in your home before embarking on an energy-efficiency program. You could save lots of cash by doing it right the first time. From your local energy-efficiency office get a leaflet that shows you how to carry out a simple home-energy audit. Check your last four gas, electricity, and oil bills to see where most of the money goes. You'll soon see that space and water heating is the most expensive, followed by light and appliances. Check to see whether you have draft proofing and insulation in the attic, on pipes, and in cavity walls. Make a priority list of what to do.

Recycle rain

Our friends in the United Kingdom have had success recycling rainwater. Now that the British water industry has been privatized, and with water metering soon to follow, all sorts of suggestions for better use of water resources have been made. One company has devised a new scheme not only to collect the 400,000 gallons (700,000 liters) of rainwater that falls on the average semidetached house but also to purify and filter it under ultraviolet light, in order to remove all traces of bacteria and viruses, so that it can be reused. The cost (at about $5,100) has a pay-back period of ten years, which may look appealing when higher water bills start coming in.

Use off-peak electricity

If you don't have gas central heating, you should choose the type of electric heater you use very carefully. Small bar-type heaters consume enormous amounts of energy and should be used as little as possible. You may be able to take advantage of off-peak electricity supplies, which are cheaper to run, as they use energy at night and store it for when you need it in the daytime.

Stop heat escaping through your floor

Gaps in floorboards can help heat to escape. You can close the gaps with wooden beading, plastic wood, newspapers, and even plaster. Several newspapers laid flat under a carpet can act as an underlay and help to prevent heat loss. They will make your carpet seem thicker, too. Wall-to-wall carpeting prevents heat loss through the floor.

Use the sun to heat your home

If you are lucky enough to have south-facing windows, you can use the sun to help heat your home. Open curtains and doors in sunny rooms, and let the gentle heat fill the rest of the house. You may find that you can turn down the heating in these rooms to save extra energy.

Fit a thermostat

A thermostat with a time programmer for your central heating could save you up to 20 percent of your present fuel bills. That in turn means that you will save energy and reduce air pollution such as acid rain and carbon dioxide, which is given off by burning fossil fuels. Set the controls to fit in with the time you get up and go to bed at night. You can always override the clock if your schedule changes unexpectedly.

Control the heat in each room

By using thermostatic radiator valves you can set the temperature in each room at the desired level. This is very energy efficient, especially if you use some rooms rarely. The valves require no electrical connections and can be fitted by any competent plumber or do-it-yourself enthusiast.

Make the heat from your radiators work for you

One simple way to make the heat from your radiators really effective is to fix ordinary kitchen foil on the wall behind them, so that the heat is reflected into the room rather than escaping through the wall. This is especially energy saving for radiators on outside walls. For radiators under windows, try building a wide windowsill, which will help reflect the heat into the room. Don't let curtains hang over radiators either; you will just encourage the heat to escape through the window. Shorter curtains are a better solution.

Insulate water tanks and pipes

If you have just insulated your attic to save energy and money, you will find that the space in the attic will be colder. It is important that you insulate all the pipes and the cold-water tank in your attic to stop them from freezing up in winter.

Get an energy-saving grant

If you aré on a low income, you could apply for a grant to insulate your home. The eligibility of a grant for home improvements changes often, and according to each local authority area, but it is worth asking your social services department. You could get financial help to save precious money.

Draft-proof your windows

As much as 25 percent of the heat in a house can be lost through drafts, and a good deal of that slips through ill-fitting windows, sashes, and locks. You can fit durable plastic strips for very little. For very narrow gaps and uneven windows use silicone rubber sealant, which is very cheap, although you may find that it cannot cope with seasonal changes easily.

Do-it-yourself double glazing

Double-glazing is expensive. Penny for penny, it does not make financial sense simply as a means to save energy, unless you are replacing your windows anyway or have already insulated the rest of the house. Double-glazing will muffle external noise and cut heat loss, but it may take up to twenty years to recoup the cost. But you can create do-it-yourself double-glazing: tack plastic sheeting into place around your windows. This is extremely cheap and effective, and its cost will be recouped within a few weeks. You will probably have to take it down in warmer weather.

Use cavity-wall insulation

Cavity-wall insulation means filling the cavity or gap between your walls with an insulating material. You will have to find an experienced contractor to do the job for you. This type of insulation is cheaper than most others, but you should avoid ureaformaldehyde (UF) foams, which can leak gases into your house, or foams containing chlorofluorocarbons, which destroy the ozone layer.

Turn down the heat

If all the households in the United States lowered their heating by about 6° F (3° C) for a twenty-four-hour period, the equivalent of 570,000 barrels of oil could be saved. It doesn't make sense to heat a home to excess. Good insulation, lighting, and thermostats will keep your house at a pleasant temperature. Don't reduce temperatures below 64–68°F (18–20°C), though, particularly if you are elderly.

Forget the chandelier

Unless you live in a stately home, you are not really going to have much call for a real chandelier, but many people do have multibulb lighting, which uses expensive and wasteful numbers of specialist bulbs. You could cut out one or two of the bulbs, but it would be far better to change to a different type of lighting and discourage the use of multibulb lights.

Write to the government about saving energy

Saving energy in the home is one of the most important ways to protect the environment. Every kilowatt hour of electricity produces 1 kilogram of carbon dioxide, which causes global warming. Every therm of gas produces more than 5 kilograms of carbon dioxide. The government publishes leaflets and distributes information about cutting energy. By writing to your congressman you can show that you care about energy efficiency as an issue.

Use only pure white spirit as a paint remover

Commercial paint removers contain methylene chloride or dichloromethane, a chlorinated hydrocarbon solvent. It is dangerous to those with heart diseases, and the caustic will burn your skin within seconds of contact. No data are available to indicate the extent of environmental damage after the solution is washed down your sink, but it has been shown to be carcinogenic in animals. These products should be used only in well-ventilated rooms and never near naked flames, as the vapors give off phosgene, which is extremely toxic. Use white spirit as a paint remover and water-based paints wherever possible.

Don't get into a sticky mess with glue!

The industrial processes used to manufacture glues and resins used by millions of do-it-yourself enthusiasts cause great concern to environmentalists. Solvents like methyl ethyl ketone,

carbon disulphide methylene chloride, cyanacrylate and epoxy
resin adhesives in general can cause pollution where they are
manufactured as well as causing dizziness, numbness, and
rashes. One case study undertaken by the London Hazards
Centre found that a bricklayer who used an epoxy resin adhe-
sive suffered badly from nerve damage in his legs and from
epoxy dermatitis (red itchy rashes) on his hands and neck. It
was months before the rash cleared up, and the man is still
suffering from circulatory problems. Do use organic glues and
adhesives, and always wear gloves.

Green up your walls

Wallpaper originated in France in the fifteenth century as a
cheap substitute for richly woven tapestries. Decorated with
stenciled or hand-painted designs, it became very popular after
the rise of paper mills around Europe. These days we buy
billions of rolls of wallpaper for our homes and offices. If you
are environmentally conscious, you may prefer to avoid wallpaper with plastic laminates and formaldehyde resins, and may
even consider painting your walls instead of using precious
paper. Painted walls are easier to keep clean and bright and do
not need to be redecorated as often. If you want to introduce
colors and patterns, stenciling is a very cheap and easily mas-
tered alternative to wallpaper.

Do not buy asbestos cement

The production of most asbestos materials is prohibited for
health reasons, but asbestos cement still slips through the net
because it is cheap and people are fooled into believing that the
fibers cannot escape into the air. It is used mostly for corru-
gated sheet roofing, walls of sheds, and internal wall panels in
apartments, cold-water storage tanks, lining and shelves in
cupboards, weather boarding, decking tiles for flat roofs and
sheeting for fire doors. Asbestos cement causes problems at
every stage of its production, use, and disposal. The water
supply may be contaminated; asbestos causes cancers in hu-

mans; people who work with it suffer from an increased risk of death from lung diseases. There is no ''safe dose.'' The erosion of buildings by acid rain is causing even more problems, as the asbestos fibers are washed out or released into the air. The Fraunnhofer Institute in Germany estimates that 1 square meter (just over 10 square feet) of asbestos cement may release 1,000 million asbestos fibers an hour. Companies producing this deadly hazard are increasing production in developing countries. Do not buy it: boycott it.

Ban wood preservatives from your house

Pesticides, used as wood preservatives, can last for twenty or thirty years, which is why you can get timber-treatment guarantees for so long. Nearly all are suspected carcinogens and some, like Dieldrin, have been banned for a long time in the United States. Pentachlorophenol (PCP) has been blamed for over 1,000 deaths worldwide; just 3 grams of Lindane is a lethal dose; and tributyle tin oxide (TBTO) is a nerve poison. These chemicals and others are sprayed regularly in our homes, and they are now widespread in the air, in water and in soil. Cancer-causing preservatives should be banned immediately, and other chemicals should be used with extreme caution, following strict safety and health guidelines.

Look after the bats in your roof

One of the main problems for bats in country areas is the spraying of chemical wood-preservative treatments in roofs, which can kill them. While no treatment can be absolutely safe, the Nature Conservancy Council suggests using zinc naphthenate because of its relatively low toxicity. Other solvents in the preservative may be harmful, however, and you should carry out work on your roof only if you are sure that there are no bats in the roof space. Contact your local nature conservation group before you start work.

Turn off lights that you are not using

It's as simple as that!

Weatherproof doors and windows

Cold air can pass through the tiniest crack. It can freeze a hall if the door or window is not draft-proofed. The energy leaking from North American windows in 1988 was estimated to be equivalent to 522,000,000 barrels of oil. You can be saving up to 10 percent of your energy bills by simple and very inexpensive draft-proof stripping.

Try pure cork for insulation

Many insulating foams use chlorofluorocarbons, which damage the ozone layer. Others contain tiny particles of urea-formaldehyde foams. Use pollution-free and safe insulation materials. Always ask what they contain. Try pure farmed cork without chemicals.

Insulate your attic

If you have not already done it, insulating your attic could be one of the quickest and most cost-effective ways of saving energy and preventing pollution. The insulation should be 6 inches (15 centimeters) thick. If you have less than this, top it up. You can usually tell which houses have poor or no insulation—just look up at the roofs on a cold day in winter and note where pigeons have settled. The heat of badly insulated houses is wasted on warming the roof tiles. Insulation could save you as much as 25 percent of your fuel bills.

Close off rooms you do not use

If you have more rooms than you need, think seriously about closing them off and switching off radiators. You could greatly reduce the area of your home that needs heating.

An organic end for flies

If flies are buzzing around your home or near your trash cans in summer, and your first reaction is to reach for some fly spray—stop! Not only are fly sprays packaged in aerosol cans but they also contain tetramethrin and phenothrin, both highly toxic to fish, aquatic life, and bees. The production of these pyrethroids is considered by the World Health Organization to be "moderately toxic," but a dose of 0.5 ounces (15 grams) has killed a child. Citrus oil repels flies just as easily, and boiled sugar, syrup, and water on strips of brown paper make a less toxic death trap. Use a fly swatter, too. Keeping your home clean is also a good fly repellent, as flies feed off grease, rubbish, and old food behind refrigerators and stoves.

Paint everything green!

Americans buy billions of dollars worth of paint each year, often without knowing the content of the cans and the environmental effects they may have. Public concern about the levels of lead in paint has ensured that only a few companies now include it, but paint contains other toxins. Asbestos, chlorinated hydrocarbons, xylene and trichloroethylene are just some of the chemicals you should avoid. For your own safety, avoid breathing paint fumes or smoking cigarettes while painting. Where possible, buy water-based or nontoxic paints, which contain safe, organic ingredients.

Rent or borrow equipment

When it comes to specialist equipment, America scores high in sales of do-it-yourself equipment that uses a great deal of electrical power and may be used only once or twice. Why not rent or borrow equipment instead? Friends and neighbors can cut down on needless consumerism by swapping their equipment.

Do not buy electric when people-power does just as well

The more specialized and elaborate your do-it-yourself tool, the more likely it is that it will use electrical energy. Lawn mowers and electric screwdrivers are just two examples of tools that do jobs that could be done by people-power instead of electrical power. You will save money, too.

Get rid of methyl chloroform

Methyl chloroform is used as a solvent in paints, adhesives, and varnishes, and it is estimated that some 544,600 tons of it were used in 1985 alone. Methyl chloroform is an ozone destroyer and has an atmospheric lifetime of 8.3 years. Because it is relatively stable, it has been substituted as a solvent for some of the faster ozone depleters, but this has meant that production has increased. Ask about methyl chloroform. Don't buy do-it-yourself products that contain it; switch to safer alternatives.

Can you breathe?

Formaldehyde is a toxic gas found in ordinary household products such as glues, chipboards, plywoods, and insulating foams. A study in the United States found that about one in five people could be sensitive to the effects of formaldehyde; the symptoms include dizziness, allergic reactions, and headaches. Increased exposure increases sensitivity, and the gas itself is highly toxic if inhaled. It has been proved to cause cancer in animals and is a suspected human carcinogen. You should avoid formaldehyde whenever you can. If you have to work with it or if it is already present in your home, keep your rooms well ventilated and wear a mask when fitting insulating fiberglass materials.

Convert your junk

Convert and restore junk to prevent pollution and safeguard resources. Make a door into an occasional table or desktop, or a wardrobe into a kitchen cupboard; swap legs and drawers; pick up junk from dumpsters and design cheap furniture yourself. There are plenty of books on the subject, and it could become an absorbing hobby.

Don't use a blowtorch

A blowtorch may be a handy way to get rid of paint on wood, but burning it off will release toxic fumes if the paint contains lead. Use a water-based paint stripper instead, and choose organic paints when repainting.

Use solar power

Solar panels may look like futuristic, expensive gimmicks, but new designs are proving all the time that it pays to install them. One model can save you up to 70 percent of your heating costs, yet costs only $2500 to install. Solar panels are clean, safe, and quiet. They have no moving parts that need maintenance, and they are made from silicon, the world's second most common element. Even in cloudy conditions they take energy from the sun, a free, renewable source that is available to everyone and causes no environmental damage. The sun is definitely one of the energy sources of the future.

Recharge your flashlight

Buy a rechargeable flashlight, and use it only in emergencies.

Clean up the drains

Use baking soda for the drains. It prevents blocked drains and stops you from putting nasty chemicals into the water system. You are bound to save on plumbers' charges, too.

Gardening

Collect rainwater to wash your car with

Car washing wastes precious and expensively treated water. Why not collect rainwater or old bathwater and use that to wash your car with? Cut down on the number of washes you give it too—you will have more time for important environmental action.

Don't always burn dead leaves and stalks

Moths and other insects spin tiny cocoons under dead plants in autumn. By burning the plants you could be destroying both valuable helpers and food for your garden. It is far better to make a compost heap and to only burn a small amount of garden refuse. Check underneath the bonfire for hedgehogs first, especially in November, when these creatures crawl under piles of leaves to hibernate.

Plant wild fruit trees

Avoid imported fruit trees. Ask a nursery for information about native varieties, and restore heritage, beauty, and wildlife to your garden.

Germinate your own seeds

Many trees will survive for longer if you germinate them from local seeds; they are more likely to be adapted to the climate and soil, and you will be conserving important local stock. The grafting or budding of fruit trees is an ancient practice, and there are plenty of books that contain instructions on how to do it.

Never use paraquat

Although restricted in most of the United States, paraquat is used in other parts of the world. Exposure to very small amounts of paraquat used as a garden pesticide can cause skin rashes, inflammation of the eyes, and the delayed healing of wounds. It is acutely toxic and has killed people even when they have spat it out. There is no known antidote.

Let the dandelions grow!

Dandelions, thistles, clover, and nettles attract butterflies. With the removal of hedgerows, the loss of many wild plants, and the increased use of pesticides, our gardens could become sanctuaries for butterflies, so set aside a section of your garden for one or all of these plants. (Dandelions are also tasty additions to salads.)

Plant nectar-sweet garden flowers

Honeysuckle, lilac, lavender, thyme, marigold, and sweet william bring color and beautiful scent to a garden. Butterflies are attracted to these flowers, too.

Make your pond safe

Garden ponds can be a death trap for small mammals and children. Ensure that small animals can escape by building slip ways of half-submerged bricks or rocks around the edges of a pond. Spread coarse netting across it to protect small children.

Do not buy peat

The vast peat deposits of Northern Canada are being used to supply gardeners in North America with peat for potting and digging. But the seemingly empty wetlands hold a rich diversity of valuable wildlife including rare species of plants, migratory birds, mammals, and insects. While Canada's wetlands seem to stretch endlessly you shouldn't be fooled. Use manure or your own peat from a compost instead. Do not buy peat for your garden.

Feed good peanuts to birds

Titmice and chickadees can eat a third of their own weight in peanuts, but many are accidentally killed by well-meaning enthusiasts, who feed these birds old and musty peanuts, which are poisoned by aflatoxins. Strict guidelines on human consumption of peanuts operate in many countries, but few people are aware of the dangers of feeding moldy peanuts to birds.

Paving crazy

Don't concrete over your garden—not even small areas. The concrete prevents vital solar energy from reaching the soil and makes the earth underneath useless. Natural drainage is hampered. Concrete is not as durable as stone in any case. If you want to construct a path or patio, use stone, bricks, wood, or slate. Encourage growth between the cracks. Check your local garden center for inexpensive and attractive ideas.

Encourage birds

Hang feeding bags on trees, or set up a bird table and encourage birds to dine there. Be sure to place the table out of reach of cats. So much of our countryside has been ruined by pesticides and chemicals, drainage and leveling that our gardens could be real havens for migrating birds.

Bonfires can be a hazard

Don't light bonfires just for the sake of it. You shouldn't burn household rubbish or wood, as these may give off toxic fumes. Consider your neighbors: have they got washing on the line, or are small children playing nearby? Check whether your bonfire will burn area trees or bushes.

Protect seedlings

Protect seedlings from weeds and insects by laying old carpet or underlay around them. Make small holes in the places where the seedlings will pop up.

Grow herbs

Fresh herbs add a certain quality to a garden and make an ordinary meal special. They are easy to grow, attract butterflies and bees, keep away slugs, and often have medicinal properties. Try lavender (my favorite) for potpourri, and chives, marigold, rosemary, marjoram, sage, and thyme to eat.

Buy organic untreated seeds

Packets of seeds were first marketed in Philadelphia in the 1880s, when Washington Atlee Burpee started offering flower and vegetable seeds as food for the poultry that he sold. People wanted the seeds more than the poultry, however, and by the early 1900s he was offering over 200 packets of vegetable and flower seeds. Today billions of packet seeds are bought all over the world. Use good producers like the Seed Savers Exchange, which sells seeds without fungicides and chemicals to help them grow. They will be much hardier.

Grow your own food

Homegrown food saves energy and eliminates waste, chemicals and artificial fertilizers (provided you do it organically, of course); it also provides majestic, mouth-watering meals. Grow beans, broccoli, soya, cabbage and cucumber, garlic and onions, lettuce and tomatoes, parsnips and even melons. And why not try some of the more exotic vegetables, which perhaps you can't afford to buy from supermarkets, like radicchio and asparagus?

Sow native flowers

Encourage native flowers grown from seed collections provided by garden lovers in your neighborhood. They will feel welcome in familiar conditions and will grow well. They will also cut down on packaging and transporting, and you'll know that they have not been sprayed with chemicals.

Attract birds

Bird groups and organizations will readily supply lists of the many flowers and shrubs that can be grown to attract a wide variety of birds into your garden. Plant cotoneaster for blackbirds and thrushes, for example, and thistles for goldfinches. Don't shoo birds away for eating fruit off the trees; they will take their part in the food chain by eating small mammals and insects galore.

Compost for city dwellers

Those of you who haven't got a garden, or who want a compost quickly, should collect your household food waste, preferably chopped as small as possible, and tie it up in a black plastic bag. Place the bag in a sunny spot for a month and you should have a good store of rotting waste, best when you can still see the shape of the original organic matter and its color is fairly dark.

Forgo the motorized mower

Lawn mowers were invented in England in 1830 by a factory worker called Edwin Budding. Larger versions of his device were pulled by horse around English stately homes (I have seen them in Delhi being pulled by cows). The motorized gas- or electricity-guzzling mower of today is an expensive and wasteful gadget for most ordinary lawns; a hand-pushed model will do the job—and the effort required may prompt even the garden-proud to leave the grass to grow!

Set aside an area for nothing but wildlife

We are so obsessed with order and tidiness in the garden that we forget that chaos can be very productive. A wild corner will soon buzz with life, given encouragement, and provide a habitat for aphid-loving insects and mammals.

Green garden furniture

In summer your garden is an extension of your house, and some furniture is essential for a relaxed and congenial atmosphere. Traditional softwood and cast-iron seats need constant painting or treatment to keep them in good condition; the hardwoods used for some garden furniture come from tropical rain forests. Check on the source of your furniture before you buy it.

Choose companion plants

Companion plants are important to any organic gardener because they repel bugs such as aphids and feed neighboring plants with trace minerals. Garlic and onions are great bug repellents, but they shouldn't be planted together. Sow onions next to carrots and garlic next to roses. Herbs deter slugs. Get organized about your garden before you start to plant and you will save yourself a lot of time weeding and thinking up nonorganic ways of getting rid of pests.

Build a greenhouse

A greenhouse is a boon: it is a warm refuge on a chilly day and a perfect place to grow your seeds and keep tools and equipment. In it you can grow and protect plants that might not survive in the garden, especially when it's cold, and it is a homemade store for solar energy.

Dig everything

One of the most effective ways to give soil a new lease on life, create good drainage and discourage harmful insects that feed on roots is to dig it. It also gets your muscles working!

Leave your lawn alone

Carole Rubin, one of Canada's top organic gardeners, sums up our attitude to manicured gardens well when she says, "If you want your lawn to look like a green rug, you're probably better off with a carpet." You must be prepared, whether growing organically or not, to put up with buttercups, daisies, and clover. Clover is useful, because it adds nitrogen to the soil, and very few other so-called weeds will actually harm the grass.

Plant wildflowers to attract butterflies

Plant buttercups, gentians, bachelor's buttons, or knapweed to attract butterflies into your garden. Butterfly weed attracts the Monarch butterfly, and Blazing star and New England asters attract all manner of beautiful creatures. Get seeds from reputable organic suppliers and follow the instructions carefully as these flowers can be slow to germinate.

Attract the praying mantis

The praying mantis is a very useful predator; it specializes in eating bugs and insects and leaves your plants alone. You can buy praying mantis larvae from good garden centers and you can attract similar beneficial insects by planting yarrow or Queen Anne's Lace.

Look after your lawn

One quarter of the earth's surface is covered by one of the earth's most important plants—grass. The grass family includes more than 7,000 different species, including corn, wheat, sugarcane, rice, millet, and barley, though grass to most of us usually means a lawn in the garden. In the United States there are over 20 million acres of lawn, and the environmental impact of the chemical fertilizers, lime, potash, and other substances that are spread on them is enormous. (It is estimated that over 1 million tons of chemicals are spread on American lawns alone.) Cut down on the chemicals and you'll reduce the pollution in your garden and protect the environment as well.

Adapt tools for the elderly

Gardening is a pleasurable experience for people of all ages, but sometimes the elderly or frail are hindered by the lack of equipment that is suitable for their needs. Don't be put off! Help the Aged says you should choose tools that are well-balanced, light, and sharp. Use household items as cheap alternatives—for example, instead of a trowel use a flour scoop, a cup or a cut-down soft-drink container with a ready-made handle. If you can't get outside, window boxes, hanging baskets and pots will give you a fine display of flowers.

Keep bees

Bees are an important part of any garden. They promote cross-pollination of plants and flowers, and some can provide you with all the honey you need to replace your sugar. Luxurious and free, honey bees are a wonderful addition to a garden. There are over thirty types of bees, though garden bees are probably the ones you'll come across most often if you don't have a hive. They fly at lower temperatures than other species and live in holes in the ground. They have a special liking for compost heaps, and they don't sting readily, so don't try to harm them. Grow kidney vetch, pickseed, foxgloves, New England asters, red vain and the bee bee tree.

Start a compost heap

Compost is a panacea for any garden. It nourishes your soil and gives it a rich texture. Nurture your compost heap with leftover food, manure, leaves, garden debris, and even newspapers (without colored inks), but don't use it until the components have rotted thoroughly. With a compost activator (you can find organic ones) you could be using your mixture within six months. A compost heap will take care of about 30 percent of your household waste, spare you from using chemical alternatives, and save energy, too.

Stick up for the earwig

Lindane is an insecticide that is used to kill earwigs. It's a persistent poison and has been blamed for the decline of bat populations; it is also especially poisonous to cats, who can't remove it from their bodies as efficiently as other animals. But the earwig is a much maligned creature, blamed for piercing eardrums and ruining prized flowers. In fact, it is a harmless vegetarian—though it can do some damage to flowers. Insecticide used on it will destroy other useful insects such as ladybugs, which eat greenflies, and will upset the garden ecosystem.

Look after garden tools

Your garden tools are valuable. Looking after them is vital for more reasons than just cost. A blunt blade will tear grass, and that will weaken it and attract weeds and disease. You can destroy roses by pruning with blunt pruning shears. Don't buy tools with handles made from rain-forest wood.

Dispose of garden chemicals carefully

If you have just decided to go organic, you may find that you have a selection of dangerous and poisonous herbicides and pesticides in your garden shed or greenhouse. The only expedient for most of us at the moment is to send them to our local councils for disposal. They should *never* be poured down a drain, as this would add more poisons to our already polluted water system.

Grow green manure

Green manure is composed of plants that are grown especially to feed the soil. They are generally fast growing and produce a mass of weed-smothering foliage. Sown in winter, when the soil is idle, they capture the microorganisms released in the autumn and store them for the spring, protecting the soil from frost at the same time. Grow mustard, rye, fenugreek, and clover.

Save water

The average North American gardener overwaters his or her lawn by up to 40 percent, which wastes over 10,000 gallons per acre. Overwatering is an easy way to destroy your lawn and use up precious summer supplies. You could save a quarter of the water used in the home each summer if you gave up watering the lawn altogether, but if you must, use sprinklers and hoses in the evening, when the water will evaporate much more slowly. Don't water lawns in droughts. Grass that is going brown will recover after rain falls again. The water is crucial for more basic needs.

Prepare your soil

Pest-resistant plants and laws need a good healthy soil. You may need to feed your garden a regular diet of organic fertilizers to keep it healthy. Test your soil, and get guidance and help from the many books and reports that are available. The importance of this cannot be overstated.

Choose organic fertilizers

Fertilizers put back into your garden what plants and the rain take out of it. They are vital for the well-kept garden, as all plants will die eventually if they are "starved" of nutrients.

The three main nutrients a garden needs are nitrogen, potassium, and phosphorus, but it may also need calcium, iron, zinc, and magnesium in small quantities. Organic fertilizers contain fish and blood meal, kelp meal and wood ash, as well as compost, your garden's richest and cheapest food if you make it yourself.

Build a bat box

Virtually all American bats are harmless, and they can devour thousands of insects in an evening. Bats are protected by law, and you can encourage them to use your garden by building a bat box. Contact your local conservation group for ideas and help.

Feed your birds well

Don't feed birds artificial food, such as bread and nuts, after March. This food isn't usually nutritious enough for young birds in spring. They will mostly feed on worms and berries during spring and on artificial foods such as nuts, apples, suet and oats (bread should be soaked) during winter. Don't feed them desiccated coconut, which swells inside them and can kill them. Encourage spring food by planting trees and shrubs, which in turn produce berries and seeds and support an insect population.

Provide shelter for birds

Birds often need some help in finding a home, especially as the numbers of trees are dwindling. Plant evergreens, holly, or a thick hedge for shelter in winter, and persuade all kinds of birds to remain in your garden by providing boxes.

Build a pond

A pond in your garden will set you apart from other gardeners. It will attract frogs and water beetles, and you will be able to watch birds around the edges and goldfish eating mosquito larvae. Use hard-wearing butyl rubber as a base for an inexpensive sheet-formed pool. Plant pond plants as oxygenators and floaters to prevent the buildup of algae.

Get rid of aphids organically

It is counterproductive to use insecticide to kill aphids. It generally kills, too, the predator animals that would otherwise get rid of these pesky insects for you. A soft-sud mixture of environmentally safe soap liquid will remove their waxy layer and make them shrivel up and die. Grow plants to attract aphid lovers, especially lady bugs. Get advice from organic garden centers.

Check out your weeders

A conventional garden guide I've just bought still suggests 2.4-D as a weed killer, as it is effective against many annual and perennial weeds. Yet evidence shows an increasing chance of developing a rare form of cancer among farm workers in the United States who use it.

Mulch it

If you don't like digging, why not mulch your garden? Mulching could increase your vegetable yields by as much as 50 percent, and it's easy to do. Just flatten weeds and flowers in autumn, cover with a thick layer of compost or well-rotted manure, and cover over with newspapers or old carpet (even black plastic bags). By spring you will find a wonderful, rich soil for planting in.

Plant holly around shrubs and bushes

Planting holly around your shrubs and bushes keeps cats away, as they don't like the prickly leaves. Birds can tolerate them quite happily, however, so offer them the protection of holly.

Grow by the moon

Sowing seeds before a new moon encourages root growth because there is a stronger gravitational pull to the earth. Trans-
planting and cutting at the time of a full moon encourages
better leaf development, and the properties of herbs and vegetables are said to be richer.

Convert your nettles

Nettles are weeds with many uses. Five hundred species are
found all over the world. Nettles in your garden are the sign of
a rich, fertile soil. Collect them and put them into an airtight
container. Cover with water and leave for four to six weeks.
Use undiluted as a pesticide and diluted with ten parts of water
as an incredibly rich fertilizer. And it's free!

Fool them with marigolds

Marigold tagetes are wonderful for fooling aphids. Their lemon
scent also stops dogs and cats from jumping on your flowering
borders. Greenfly are completely fooled, which could save the
rest of your flowers.

Elderberry or rhubarb mix as a weapon against aphids

About 4 pounds (1.8 kilograms) of elderberry or rhubarb leaves
boiled with 1 gallon (4.5 liters) of water for one and a half
hours make a superb organic pesticide. Add a teaspoon of
ordinary soft soap, so the mixture will remain on leaves, and
use as a thin liquid spray against aphids.

Beer gardens work

Half fill an old tin can with beer and dig it into the ground.
Slugs will be instantly attracted to this treat, and you'll be able
to collect dozens of them in the morning without having to
resort to chemicals at all.

Root it organically

To promote root growth in plants without using fungicides, why not use an ancient organic method? Cut branches of willow and steep them in water for between six and twenty days. The resulting mixture is an organic rooting hormone, excellent for fuchsias or similar hardwood plants. You have only to dip the cuttings in to promote root growth. The mixture is deadly, so keep children and pets away and label very carefully.

Plant nasturtiums

Climbing nasturtiums have edible flowers but are also useful if you want to attract blackflies away from your cucumbers. These pests will be attracted to the back of the nasturtium leaves, so your vegetables can enjoy some peace.

Do not plant rhododendron

The rhododendron's growth speed is phenomenal. Ten thousand seeds weigh only 0.04 ounce (1 gram) and are quickly dispersed by even light winds. The bush does not support any insect life, prevents woodland regeneration, is poisonous to mammals, and produces an acid humus that stops regrowth even if the plant is removed.

Pets and the Environment

Use pennyroyal to get rid of fleas

Pennyroyal is a member of the mint family and was first used by the Romans to drive away fleas. You can use dried pennyroyal leaves packed in your cat's or dog's collar, or wash your pet frequently in pennyroyal oil to ward off these annoying insects. Don't go to the expense of using artificially created fleabanes, which often contain toxic organophosphate compounds and are the cause of most fatal cases of poisoning among pets.

Don't buy exotic birds

Europe and North America are the biggest markets for wild birds. Poorer countries are encouraged to catch and sell millions of parrots, macaws, cockatoos, flamingos, budgerigars, and birds of prey. Some are endangered species, and nine out of ten die before they reach their destination. They are crammed into crates in plastic tubes, without adequate food or water. Their own body heat often suffocates them. Many others die waiting in quarantine. These exotic birds are not pets; they should be allowed to fly about in their natural habitat without persecution.

Oppose the giving of live animals as prizes

At fairgrounds, bazaars, and carnivals prizes often include live pets such as goldfish. These fish are stressed and suffer from oxygen starvation, the main cause of their death. A minute change in water temperature can kill a goldfish instantly, and tipping them out of polythene bags into bowls of tap water when you get home can be fatal. They need room-temperature water and should never be kept in small goldfish bowls, which do not provide them with the oxygen they need. Don't accept a goldfish as a prize, and encourage fairground traders to adopt ideas that are more ecologically sound.

Keep your pony or horse in the countryside

A pony is the second most popular pet for a young child, whether in the countryside or town. The ASPCA reports a worrying trend in the practice of keeping ponies in urban areas. They need to have at least 2.5 acres of land and a stable.

Don't buy a throwaway pet

Each Christmas kennels take in more and more pets—unwanted gifts. Strays and those left at rescue homes, however, account for only a small percentage of the animals (especially dogs) that roam the streets. If you think you want a pet, consider the matter carefully, and buy one only if you know that you will be able to give it time for exercise and attention. Over 20 million dogs and cats were put down in the United States last year. Don't give other people pets as presents unless you are certain of their commitment.

Train your dog to use the gutter

A puppy is old enough to be house-trained at eight weeks old, and it usually takes about a month for it to be fully trained. It wouldn't take much longer to train a dog to use the gutter—yet a surprising number of dog owners don't seem to be concerned about foul pavements. Clean up the streets. Train your dog.

Use garlic

Garlic is an important food for human beings, but it can be useful for animals, too. Pet lovers recommend garlic as a non-toxic and safe alternative to chemical tablets for getting rid of worms in cats and dogs. Mix raw garlic with their food.

Don't let your pet annoy the neighbors

Keeping a dog in the city can be a problem for the dog, which may not get out as much as it needs to, and for the neighbors, who have to put up with barking. If you have a dog, you have a responsibility to look after it and stop this annoying form of noise pollution whenever possible. Plenty of walks and exercise should keep your dog happy.

Consider hens for your garden

Keeping hens in your garden, if it's big enough, is a wonderful alternative to dogs and cats. They can be fed on grains and scraps from the kitchen and will produce fresh organic eggs for you as well as manure for your garden. Roosters can be noisy, however, if you are not in the country. Think carefully before getting one.

Say no to the tropical aquarium

Tropical fish come mostly from freshwater sources in tropical countries and are exported for sale around the world. They are difficult to breed in captivity, and about 60 percent of imported tropical fish die within a year of purchase. Tanks for these fish are often too small, and chlorinated water affects them. Exotic aquatic life should be left in its real home.

Give your pets a healthy diet

Dogs and cats today eat a very unhealthy diet. The canned food that we buy them contains sugars, stimulants and preservatives to which the animals may become addicted. The cans may be contaminated with lead, and the meat inside is unfit for human consumption, which is why it's fed to our pets. Dogs and cats should ideally be fed on the raw organs of their prey. Although they will hate the diet at first, a raw-fed animal will be far healthier in the long run. But such food is almost impossible to procure, so careful thought is necessary when it comes to planning a pet's diet. Although cats need about three-quarters of their diet to be meat, dogs can survive on just one-third meat, the balance to be made up by grains, scraps, and vegetables. Alternatively, dogs can be raised as vegetarians from birth; vegetarian organizations will give you advice about suitable foods. Cats and dogs seem to suffer less from parasites when on a healthy diet, so you may find vet bills cheaper, too.

Put a bell on it!

Cats are responsible for killing millions of birds each year. Help to reduce these deaths by putting a bell around their necks, the louder the better. If possible, in summer and during the breeding season don't let your cat out between first light and 8:30 A.M. and at dusk. Parent birds are busy finding food for their fledglings and are more distracted at these times.

PERSONAL
GREENING

Hygiene and Beauty

Be assertive

When we shop for beauty products in large department stores, we're often approached by well-groomed sales assistants in clinical white coats, who look like trained beauticians, dermatologists, or chemists. They are not. They are there purely to sell you goods. At least one-third of the beauty products and treatments bought from these assistants are purchased because customers are not assertive enough to say no. We are buying products we don't really need, wasting resources, and energy.

Make your own hair spray

Modern hair sprays come in aerosols and contain a wide variety of synthetic chemicals, including cancer-causing polyvinylpyrrolidone plastic (PVP), formaldehyde, artificial perfumes, and alcohol. Hair sprays are extremely flammable and often carry warning labels about the consequences of accidental inhalation or spraying into eyes. If you really must use hair spray, why not try making your own? Our recipe suggests using whole lemons chopped and boiled in water until half the mixture is left. Strained and cooled, and applied to clean hair, it can make an effective alternative to manufactured sprays. Refrigerate the mixture in a pump-action container for a cheap, safe, nontoxic spray.

Banish the polish

Nail polish and nail-polish remover are subtle mixtures of toxic chemicals. Phenol, toluene, formaldehyde and xylene are the main ingredients. Prolonged exposure to chemicals such as xylene affects the nervous system, and chemical reference books suggest avoiding skin contact. Healthy nails need contact with the air. You'll have to banish the polish. Sorry: there are no green alternatives.

Go for an ammonia-free permanent wave

Hair-perm solutions generally include ammonium thioglycolate (or thioglycollate), which is highly toxic if you swallow or inhale it. It causes rashes and swelling, and it can be absorbed through the scalp into the bloodstream. The ammonia can also make breathing difficult, so look for an ammonia-free product and ask your hairdresser to use it on your hair.

Dye your hair green

Hair colorings are generally extremely dangerous. Some are known carcinogens, although loopholes in American regulations prevent the Food and Drug Administration from issuing warnings about them. They contain ammonia, detergents, ethanol, glycols, colors and fragrances, hydrogen peroxide, lead, and sulphur compounds. Unless you are absolutely certain about their contents, it is better to stick to natural substitutes like henna or paprika for red hair; ginger, hot coffee or nutmeg for brown; chamomile or ceylon tea for honey.

Crunch a clove for fresh breath

Don't be fooled into thinking you need a mouthwash containing ammonia, hydrogen peroxide, ethanol and formaldehyde to make your breath fresh. This is not the answer to your problem. Try crunching a clove, brushing your teeth, and changing your diet first. If that fails, see a dentist.

Buy cruelty-free

Every year over 20 million animals die agonizing deaths in tests conducted in the cause of profit around the world. New beauty products, perfumes, lipsticks, nail polishes, talcum powder, hair spray, and eye shadows are tested on dogs, rabbits, chimps, and rats. Pellets of lipstick are force-fed to laboratory animals to test lethal dosages. Shampoos and lotions are dripped into rabbits' eyes to test their effects, and, as rabbits cannot blink or cry, they are unable to restrict the pain they have to endure. There are hundreds of cruelty-free beauty products on the market today, so there is no longer any need to buy cosmetics that have been tested on animals.

Say no to musk oil from deer

Glands in the male musk deer secrete one of the most valuable fluids in the world—musk oil. To obtain 2.2 pounds (1 kilo) of musk, forty adult males have to be killed. In 1986 it was reckoned that 37,000 of these endangered animals were killed worldwide, mostly illegally. Because of the scarcity of the deer, more than 300 synthetic chemicals have now replaced the real thing. Animal rights groups say that some perfumes still contain deer musk, including Chanel No. 5, Guerlain's L'Heure Bleue and Rochas' Madame Rochas. International legislation protects the musk deer, but evidence from the World Wildlife Fund shows that Japan and France are still importing oil, about 80 percent of which is thought to be illegally obtained.

Do you need a moisturizer?

The skin-care market is worth an estimated $2 billion a year in the United States and is growing fast. Prices for moisturizers range from $.85 to $85 for 1 ounce (28 grams). Most dermatologists insist that there is little difference between them. One of the world's leading dermatologists, Professor Albert Kligman, suggests that you should use only two products, petroleum jelly and a good sunblock. Don't waste your money on packaging.

Ignore the hype

Forget the wild claims made for creams and lotions that will change your skin, repair DNA cells, remove wrinkles, and make you look ten years younger. We spend millions a year on beauty products, and most of that money is claimed by advertising and marketing, very little by the product itself. The marketing relies on pseudo-scientific language that proves nothing at all about the benefits of the products. Ignore the hype. Look for alternatives that don't cost a fortune (try wheat-germ oil for wrinkles, for example).

Don't use lead-based eye makeup

Surma, an Asian eye makeup that is usually lead-based, is poisonous, especially to children. In 1985 the British government issued strong health warnings about it, but most of it is imported through friends and relatives, not through shops. The continued use of lead in makeup is an environmental problem as well as a serious health risk.

Wear greener lipstick

Lipstick is an oily product based on lanolin, petroleum jelly, silicone wax, castor oil, and perfume, as well as color pigments, usually eosin. The colors known as FD and C Red No. 2 are banned in the United States but not in other countries. Titanium dioxide is a white pigment used to make pink and opaque shades. The continued manufacture of titanium dioxide has provoked campaigns by Greenpeace, as the waste effluent causes serious environmental problems. Buy lipstick without the dangerous chemicals and always cruelty-free.

Buy cruelty-free eye makeup

Most eye makeups, mascaras, eye shadows, eyeliners and eye pencils are tested on animals unless the label says otherwise. The Draize eye test is particularly cruel, as it is used on animals that can't cry or blink. Some eye makeups even contain animal ingredients. If you want to avoid these, go for products from the Body Shop or Beauty Without Cruelty, for example.

Avoid skin whiteners

In the 1960s the cosmetics and perfume industries identified a new market: catering to people with black skin. One of the more alarming products that came onto the market was mercury-based skin-lightening soap. This can cause severe liver and kidney trouble, not to mention the environmental problems arising from continued use of mercury.

Reduce your use of perfume

The world spends an amazing $2 billion on fragrances every year, which makes the perfume industry one of the biggest beauty industries in the world. We pay most of that money for packaging and advertising. Parfums Internationaux spent $10 million on promoting its Passion perfume. Every year about fifty new brands come onto the market, mostly after extravagant launches. Perfume is the main ingredient in many products that cause rashes and allergies. Some scents are photosensitive (react to sunlight) and cause irritation when worn in the sunlight. But perfume can have an uplifting psychological effect on us: why not try perfume from refillable bottles, not tested on animals, or pure essential oils?

Cut out the coral

Coral jewelry looks beautiful, but the world's coral reefs are being endangered by overfishing, waste dumping, and the trade in trinkets, including bracelets, necklaces and rings. Coral reefs support one-third of all fish species: it is their home that is being slowly destroyed around the world. Cut out the coral.

Perspire with nature

Ninety-two percent of American women and 86 percent of American men use deodorants. The vast and profitable business of producing alternatives to our body odor is quite out of proportion to the number of people who have real body odor. Antiperspirants and deodorants contain substances such as aluminum chlorohydrate, formaldehyde, and ammonia, and chemically produced perfumes and fragrances are also included. They are usually packaged in that wasteful tin, the aerosol. Aluminium is probably the most dangerous ingredient of these products. In antiperspirants it works by blocking the pores in the skin to prevent sweating. Deodorants do not stop sweat: they change the smell. Many people have suggested baking soda as the most effective alternative, but regular washing helps considerably. If your body odor is offensive, you might try looking at your diet or your life-style. Sweating is the body's way of expelling poisons; you may have to look into your intake of these poisons, especially in your diet.

Say no to badger bristles for shaving

Badger bristles are regarded as the best bristle for shaving brushes. There are many good alternatives that last just as long. We do not need to use badgers to get a clean shave.

Wind up your watch

Instead of using battery-operated watches and clocks, why not use an old-fashioned windup? It is much cheaper, and its simply technology means that it will last for years.

Soften your nails naturally

To soften the cuticles of your nails use olive oil instead of synthetic chemical formulas. It needs to be applied only once a week.

Clothes

Buy well-fitting shoes that will last

Believer it or not, 45 percent of North American women wear uncomfortable shoes for the sake of fashion, and at least 20 percent of North American men do the same. Personal expenditure on footwear in the United States amounts to $109 per person, per year, totaling some $27 billion per year. Do yourself a favor: insist on shoes that fit your feet, not the latest fashion.

Do less ironing

Irons were used by the Greeks as long ago as the fourth century B.C., but it was not until 1882 that the first electric iron was invented. It was used by the ever-growing number of households in North America that had access to electric light fittings, the sockets of which could be used as plugs. Ironing uses a great deal of energy, which in turn produces carbon dioxide, adding to the greenhouse effect. Most clothes do not need ironing at all (some people still iron their underwear and hankies!). Save yourself a lot of bother—keep ironing to a minimum. Avoid heating and reheating the iron by planning your ironing sessions. Steam irons use much more energy than dry irons.

Wear more clothes when it gets colder

There are people who insist on wearing summer clothes throughout the year and turn up the central heating in winter. This costs energy, increases pollution, and is more expensive.

In cold weather you should wear warm clothing that retains your body heat instead of trying to create warmth artificially.

Do not use mothballs

Mothballs are made with 1,4 dichlorobenzene or paradichlorobenzene or naphthalene, which are highly toxic if swallowed. Studies have found cancers caused by these substances in laboratory animals. They are extremely persistent in the environment, and long-term exposure to them can cause kidney and liver damage. Using mothballs means that you spread the vapors and particles of this chemical through your clothes continuously. Try lavender bags for a safer, cheaper alternative. Keep your clothes clean to stop the buildup of moth eggs.

Dry clothes on the line

The ultimate clothes dryer and sterilizer is the sun. If you have a garden or balcony, by far the best way to dry clothes is on a line outside. A tumble dryer uses 3 kwh of electricity to dry 8.8 pounds (4 kilograms) of clothing, which costs money and causes extra pollution.

Buying shoes from rain forests?

Some of the leather used to manufacture the 20 billion pairs of leather shoes imported annually by the United States comes from South America. Leather is one of the by-products of the cattle-ranching industry, which is destroying vast areas of tropical rain forest. Although South American countries desperately need the income to pay back their foreign debt, take a closer look next time you buy shoes, and go for shoes made closer to home.

Use alternative bleach

The sun is our most ecologically friendly bleaching agent. It also sterilizes. But if the sun is not shining, you can use sodium percarbonate, which is made from salt, limestone, and hydrogen peroxide, in preference to chlorinated bleaches. Skip the optical brighteners in soap powders; they simply fool you into thinking that something is cleaner than it is. Sodium percarborate works only at high temperatures. Use a bleach in your wash only when you really need it, and choose a safer alternative.

Use a greener fabric softener

Fabric softeners were invented purely to reduce the static that builds up on synthetic fibers. Clever marketing and advertising have led us to think that we need them for every item of clothing. They leave on all fabrics a residue of chemicals that has caused allergic reactions in many people. The softeners also break down the natural oils in some clothing and help them to get dirtier. Try using a quarter of a cup of white vinegar in the final rinse if you really need a fabric softener.

Remove spots with soap flakes

Most spots and stains can be eased off with a good scrub with soft soap flakes and warm water before you put the clothes in the wash. Do not spend money on harmful and wasteful chemicals.

Get your own spray starch

Aerosol cans of spray starch, which adds body to shirts, are generally highly toxic products. They may contain formaldehyde, pentachlorophenol, and phenols and are corrosive and dangerous if they touch the skin. They also give off toxic fumes when heated. Try dissolving one tablespoon of cornstarch in 1 pint (1.1 liters) of water, shake and spray from a refillable spray bottle.

Avoid dry-cleaning

Solvents used for dry-cleaning are known to cause health and environmental problems. Perchloroethylene is irritating to the eyes and throat in concentrated amounts and gives off toxic fumes when heated. It is toxic in water and should never be disposed of down the sink. Some solvents are known to damage the ozone layer and should be banned completely. You can minimize the risk by not buying dry-clean-only garments.

Be ruthless about clothes shopping

We all have clothes we don't really want stuffed away in wardrobes and drawers, unworn and unwanted. Be ruthless when you go shopping for clothes. It doesn't make sense to buy clothes that don't fit (when exactly are you going to lose/gain that weight?). Don't buy clothes you don't need just because they are on sale.

What's on your legs?

Women have been wearing stockings since A.D. 600, when soft leather *sykhos* covered their legs in Greece. The latinized *soccus* became the forerunner of the "sock" and was brought from Rome to the British Isles. Early British stockings were made of silk, but these days some 500 million pairs of tights per year are made of nylon. The nylon stocking was invented by the chemical company Du Pont in the United States in 1938, and in the first year 3 million pairs were sold. They did seem to last longer than silk at first, but nowadays the average woman uses twenty-five pairs of nylon tights per year. They are not biodegradable and are often linked to health problems like thrush. Wear cotton tights or socks more often, or enjoy bare legs.

Phase out fashion

High-fashion clothes are produced four times a year. Colors, trends, shapes and sizes are altered to get you to update your wardrobe continuously in line with the latest fashion. Fashion is an important part of our culture, but it has become a hyped marketing strategy to persuade naive customers to buy billions of items every year. Classic clothes that are well made stay in fashion longer because they don't need fads to sell them. Don't buy just for fashion's sake.

Buy an energy-efficient washing machine

The energy needed to wash clothes produces millions of tons of carbon dioxide, the main greenhouse gas. If you have to buy a new washing machine, choose one that is energy efficient. If we all did this, we could reduce the carbon dioxide levels produced by over 60 percent. The US government has standardized energy labeling on all electric goods, and customers can calculate savings before they buy them. Most companies produce detailed information about their models, but you have to beg showroom assistants to part with it! Do your homework, and you could save a lot of money and cut down on pollution.

You don't need animal-tested clothes

The clothing industry uses animals in a subtle way. The production and processing of fur, leather, wool, cotton, and synthetics requires the use of animal-tested chemicals, animal-tested drugs and vaccines and animal-tested bleaches, dyes and pesticides. Say no to animal-tested clothing. Watch for the growing number of businesses that can guarantee animal-free clothes.

Oppose the fur trade

Fur coats are luxury garments for those of us who live in urban areas. They are not needed for survival, and they act only as status symbols. Inhumane practices abound, including leaving animals such as the red fox, the lynx, the leopard, and the ocelot with legs caught in steel-jawed traps that have been condemned worldwide as a cruel and primitive means of capturing them. Many species are becoming extinct, but until we oppose the fur trade they will continue to be captured. The only creature that has the right to own a real fur coat is the animal itself.

Make your own clothes

Be creative and make your own clothes. If you have access to a sewing machine, the time, and some patterns, you can save money, become more self-reliant and dress more individually.

Read the labels

Reading the labels on your food isn't good enough. You must read the labels on your clothes, too. Look out for mixes of natural and synthetic fibers and washing or cleaning instructions (does the garment need to be dry-cleaned or hand-washed?). Select clothes that will cut down on energy, expenditure on chemicals, and specialist cleaners and, of course, pollution.

Buy alternative silk

China has been the world's largest silk producer for over 4,000 years, and the use of the mulberry silkmoth has become more intensive and mechanized. Fifty moths can deposit more than 20,000 eggs, which produce their silk in just three weeks. The pupa is killed in boiling water, acid, or by gassing which keeps the cocoons intact and insures the fine yarn of consistent strength that is needed for modern power looms. Hand-reeled raw silk is uneven and unsuitable for power looms, and the threads are drawn from cocoons softened in hot water. Alternatives to silk include alginate fibers from seaweed, ardil from peanuts, and vicara from maize.

Polish your shoes

Shoe polish helps to keep your shoes in good condition, so that they will last longer than just a few moments in the rain. Keep your shoes in good condition in a nontoxic way. Tests in the United States have found seven toxic ingredients in ordinary shoe polish, including methylene chloride and nitrobenzene. Methylene chloride (also known as dichloromethane) is a known cancer-causing agent in animals and as a strong skin irritant in humans. Buy nontoxic shoe cleaners that contain organic waxes, plant oils, and pigments that aren't tested on animals, and help to reduce the number of needless chemicals that we produce.

Reuse your clothing

If you don't feel that your old clothing is good enough to recycle in a charity shop, recycle them at home. Cut worn out towels into small squares for facecloths. Turn old sheets and shirts into polishers and dishcloths. I remove buttons, zippers and belts and keep them in a tin for emergencies and for when I feel creative.

Health and Well-being

Campaign against the car

The effects of the car on our health are serious. Vast quantities of sulphur dioxide, nitrogen dioxide, lead, carbon monoxide, and hydrocarbons are released, contributing to 30,000 extra deaths in the United States every year. Pregnant women and small children are especially at risk. World Health Organization limits for carbon monoxide emissions have been breached in London since 1987. As the ozone levels in the environment increase, so do hospital admissions for heart diseases. Campaign against the car for better health. Use public transportation, cycle or find other alternatives.

Protect yourself from do-it-yourself materials

In the race to save money and energy, many people are using potentially harmful materials in home-insulation applications. You should exercise extreme care, wear a mask and thick rubber gloves, and never expose your skin to the material. You should put the mask in the trash can when you have finished the task, wash all clothing, and throw away any spare insulation material, in a separate bag, immediately. Keep attic doors closed to prevent particles from blowing into the house. These warnings should be taken seriously by everyone. The business of saving energy is important, but your own health is also vital.

Don't use a douche or vaginal deodorant

Female douches generally contain ammonia, detergents, artificial fragrances, and phenol, which is easily absorbed by the skin and is very toxic. So-called "feminine hygiene sprays" are, of course, aerosols with artificial chemicals in them. Regular washing should keep you clean without these unnecessary products. If you have vaginal smells or discharges that worry you, it is far better to see a doctor than to try to cover up the smell. Try bathing with a few drops of tea tree and lavender oil.

Learn to love a tree

Trees stand proud as the most magnificent and one of the oldest of all plantlife on this planet. Some alternative health specialists actually recommend hugging trees to get energized. Five minutes sitting under an oak can make you feel serene and calm. Why not learn to love a tree?

Clean wounds with clean spirits

Some medical disinfectants—indeed, some of the best known—contain trichlorophenol, a mixture of chlorinated phenols that are known to be corrosive and dangerous. One and a half grams of phenol have caused death, and it is known to be poisonous through skin absorption. Rinsing a minor wound with soap and hot water, salt water, or calendula lotion is adequate to kill bacteria; serious wounds should be treated in a hospital.

Get rid of a sore throat with greens

I once had a cut on the roof of my mouth. I spent two weeks using chemical mouthwashes and antiseptics to no avail, when a friend told me I was being too harsh on myself. She suggested gargling or drinking boiled sage tea. It worked so well that I have recommended it for sore throats, tonsils, acne, and cuts, and it has never let me down. It tastes *awful*, so to get a child to drink it I recommend adding a little honey. Cut down on the chemicals in your bathroom cabinet, and get some sage.

Plant trees in hospital grounds

There is some evidence that patients in hospitals have a faster rate of recovery if they can see trees from their windows. Local conservation groups should work together with hospitals to create an atmosphere of beauty and peace around hospital grounds. Orchards will be especially beneficial, offering not only trees but also their fruit, and may even provide a form of therapy for patients who are well enough to help look after them.

Stick up for wild chimps

Animal experiments for medical research are a controversial issue. Many pills and cures have been tested on animals to check safety and reactions before being given to humans. There is important argument against these experiments, however. West African chimpanzees are increasingly under threat because they are being exported for medical research. Chimps have blood groups that are similar to those found in humans and are therefore very valuable for such research. The World Wide Fund for Nature estimates that 34 percent of the primate species are vulnerable. However great our need for cures, we cannot condemn other primates to possible extinction in order to cure ourselves. Not all experiments can be justified as necessary to save human lives. Chimps have been electrocuted, starved, beaten, frozen, burned, blinded, and killed, often without painkillers, for causes that are at best ridiculous and at worst barbaric.

Don't support drugs and animal experiments

In 1984, at the University of Texas, kittens were injected with the hallucinogenic drug LSD to study their behavior. The kittens became uncoordinated and displayed reactions that included vomiting and body spasms. We do not need to inject animals with such drugs to determine reactions that may be different to humans' in any case.

Cut down on drugs

Only one-third of all prescriptions are prepared and taken properly. The pharmaceutical industry tries to get us to buy more and more drugs, not just to make us better but so it can make a profit. Our life-style, the food we eat, and the stress our bodies suffer are largely preventable. Think carefully before you take tablets. Are they really necessary, or is there another way of solving your problem?

Don't eat more than you need

One of the greatest tragedies of so-called "developed" societies is that our affluence is literally killing us. A diet high in animal fat, cholesterol-rich dairy foods, excessive intakes of sugar, and refined processed food means that we suffer from heart attacks, obesity, and high blood pressure. It is one of the major causes of death in the United States. By eating less and better food you will enjoy a healthier life and help the planet by cutting down on waste, overrefined foods and packaging.

Reject animal testing to find cancer cures

One of the medical profession's best myths is that we must test drugs on animals to find a cure for that terrible disease, cancer. The World Health Organization says that 80 percent of cancers are related to life-style and environmental factors and are, therefore, largely preventable. Industrial chemicals, smoking, food additives, and bad diets are all known causes. After fifty years of experiments on animals, scientists have found that other methods of research are more cost-effective, more humane, more accurate. Clinical research on humans was the method used for the first major breakthrough for the drugs used in chemotherapy treatment. Ten rare cancers can now be cured with drugs developed by human studies. The National Cancer Institute has tested "cures" on 500,000 animals to date, with a success rate of only 0.0001 percent. This slaughter cannot be justified. Support cancer charities that do not use animals. Ask others to stop experimenting on animals.

Look after yourself

Look after yourself. Take care of your own health and you will have the energy to care for others. If you really want to save the planet, start by saving yourself first: no one else is as interested! Get enough sleep; eat a good diet; reduce your stress levels; do not expose yourself to unnecessary chemicals; and keep fit.

Think deep ecology

In the rain forests of Australia groups of people met to discuss their despair over the state of the planet. They were frightened, angry, and confused. Everything around them seemed to be in a state of collapse, and the forest they wanted to save was soon to be destroyed and replaced by a new road. From this meeting came the Council of All Beings, a unique consolation for those who feel despondent about the monumental task ahead. Many people feel that we are inseparable from nature, but few of us have the chance to enjoy an intimate relationship with it and to gain strength from the contact. Deep ecology aims to help us remember our connection with the planet, reassess our lives, and find clarity, inspiration, and commitment.

Eat less beef

Growing numbers of cattle are providing the planet with an unusual problem—they belch and break wind, giving off methane, a potent greenhouse gas. Methane accounts for 18 percent of the greenhouse effect. And that is not all. It takes more than twenty times the energy you get from your steak to feed the animal in the first place. This means that huge amounts of grain and soya, which could be used as a stable diet for hundreds of thousands of people, are grown to feed cattle.

Get rid of your own headache

Nine out of ten headaches are the result of stress, including anxiety, depression, worry, and other emotional problems. American Food and Drug investigators have concluded that tablets provide no relief for this type of headache. They recommend taking tablets only for a fever or a hangover. For natural headache relief, drink an herb tea like chamomile, mint or rosemary, get a massage, relax your neck, try to cut down the stress, and get some sleep. Sit quietly, upright, and with your feet firmly on the ground, and imagine energy coming away from the head. These simple measures could save a great deal of the money we spend every year on drugs and free resources for research into more acute illness.

Support population planning

Women's powerlessness, the degradation of the environment, social inadequacies, and hunger are the main causes of the population problems this planet now faces. For depressed countries the only source of power for the poor is children. The single most important step to be taken by First World countries is to give women the world over the power to plan their families. International policy has to change. The answer is not just to dish out drugs or injections but to share our wealth and our knowledge with everyone.

Balance your diet

An average American consumes 4.4 pounds (2 kilograms) of food every day, but Belgium was given the honor of being the leading calorie-consuming nation in 1989 with an average daily consumption of 3,850 calories. We can enjoy a delicious and nutritious diet without eating to excess. Saving this planet is about saving people, too.

Overcome your addictions

Large fertile areas of land are devoted to feeding the addictions of a few, mostly in the industrialized world, with cigarettes, heroin, sugar, coffee, and alcohol. Help is available from groups such as Alcoholics Anonymous, but awareness of the causes and consequences of addiction, in oneself and in others, is essential. We need to share responsibility globally for the damage done to the environment and our life chances by addictions.

Don't use tampons

Tampons have been surrounded by controversy since the early 1980s, when it was discovered that there was a direct link between Toxic Shock Syndrome (TSS) and tampon use in the United States. But as long ago as 1942 doctors expressed concern about their use, and more recently studies have found that up to 75 percent of women using tampons have encountered problems. Tampons also do environmental damage. The cotton is produced with the help of pesticides, and the rayon, derived from wood pulp, is manufactured with the help of chlorine, which forms dioxins as by-products and pollutes the rivers and seas. Tampons are not sterilized: the Women's Environmental Network exposed the industry during 1989 with the publication of "The Sanitary Protection Scandal." Use sanitary pads instead.

Find out about allergies

The heavily polluted environment of today is directly responsible for an increase in allergies and similar illnesses. Allergies can cause asthma (which can kill), eczema, hyperactivity, hay fever, colics, etc. Humankind has become allergic to a number of foods and products because of sensitivity to chemicals. By reducing or eliminating food additives, atmospheric pollution, pesticides, water pollution, and other chemical stimuli we can reduce the risk of allergies.

Use herbs for minor illnesses

Many of the illnesses that we suffer are minor: sore throats, acne, insect bites, and headaches. There are numerous alternatives to the billions of tablets that we put in our mouths every day, and they have been around for centuries. Use natural cures for minor illnesses. Consult good alternative health books, and buy your herbs from reputable health food shops, but, above all, get to know your body and the causes of your discomfort. You will be able to help yourself get better with a positive attitude to your body.

Use homeopathy

Homeopathy works on the principle that like cures like. Homeopathic remedies are made from herbs, salts, minerals and, in some cases, diseased tissue in minute dilutions. The potentized remedies (as the diluted medicines are called) are given to the patient to stimulate the individual's own healing process of the body, rather like vaccines. It is unwise to try to treat yourself for anything other than minor ailments, as the remedies are prescribed precisely for the individual, taking family history, likes and dislikes, and emotions into account. Homeopathy does work. I have had it myself, with excellent results. It could well be the medicine of the future, as it has little or no detrimental effect on the environment, offers a holistic attitude to patient care, and has no history of animal experimentation.

Stimulate your senses the natural way

Use essential oils to stimulate and to cleanse. They offer an alternative to aerosol air fresheners and chemical bath oils, and they are good for massages. They are derived from aromatic plants and trees and smell wonderful if warmed over a candle to give a room an extra special atmosphere. Essential oils are widely available.

Try *acupuncture*

Acupuncture has been practiced for over 3,000 years in the East. The basis of this healing art is that health is the balance between two opposing forces, Yin and Yang, and our body, in a state of constant flux between the two forces, becomes ill when they are no longer in harmony. Acupuncture rights the balance by stimulating points along lines of the body with fine needles, which conduct vital energy. It is known as a "holistic" healing method; it considers stress and emotions along with physical illness, and it is said to be 60 percent effective for chronic pain sufferers. It involves no use of chemicals or drugs and has no adverse environmental effects.

Relax!

Stress and stress-related illnesses account for a large proportion of modern sickness. Meditation may help you relax. Join a class, or try sitting quietly at home and concentrating on your breathing for half an hour.

Try *reusable incontinence pads*

About one in four women suffer from incontinence at some time during her life, and in the last few years companies have been marketing disposable paper incontinence pads, which are generally bulky and expensive. Their environmental effects range from pollution at the paper mill, in the form of dioxins and furans when the paper is bleached, to disposal. The plastic backing is not biodegradable and, when wrapped up and put in the trash can, will be around for about 500 years. Try using washable incontinence pads, which save money and offer more dignity than a bulky paper diaper.

Nose for problems

The facial tissue industry is worth $874 million each year in the United States. Bleached white tissue, which is impossible to recycle, is an unacceptable indulgence. If every person in the world used a box of tissues every month, there would be no trees left on the entire planet. Handkerchiefs originated in France in the fifteenth century, when sailors returned from the Orient with lightweight linen *couvrechefs* (coverings for the head). The anglicized word handkerchief was adopted when these cloths began to be carried in the hand. They then became a fashion item, and in 1530 they were recommended as nose wipers by Erasmus of Rotterdam. Cotton handkerchiefs are now widely available. As well as being cheaper in the long run, they are much less wasteful.

Stop smoking

Cigarettes are a hazard not only to your health but also to the environment. The poisons in cigarettes include heavy metals, dioxins, and pesticide residues, and the trillions of cigarettes smoked around the world every year create pollution for everyone. Not only does the tobacco plant take minerals from the soil and require large amounts of pesticides, but also it is estimated that 4,800 square miles (12,000 square kilometers) of forest are needed to fuel tobacco-curing factories each year. Do not harm yourself and others with cigarettes.

Avoid unnecessary X rays

By far and away the largest source of artificially created radiation that we expose ourselves to is X rays. The benefits may outweigh the risk. Serious accident victims need X rays, but you do not need them every time you visit the dentist or hospital, so take extra care to be protected and agree to X rays only if they are vital. If you do need an X ray insist on a lead shield to protect your organs.

Save the rhino

The rhino is seriously threatened with extinction. It has been hunted for over 1,000 years, and in 1987 there were only a few thousand of the species left. It is literally being hunted to death by man because of an ancient superstition that rhino horn has great magical and medicinal powers. Poaching gangs are sophisticated and well paid, often causing havoc in national parks and game reserves. Within fifteen years the rhino will be extinct unless we demand firm international action. Don't buy this type of "medicine."

Get rid of smog

One of the most upsetting things I learned early on as an environmentalist was that some of the nicest-looking sunsets are actually caused by the reaction of sunlight with chemical compounds that we have discharged into the atmosphere. These come from car pollution, factories, and gases. They can be particularly poisonous to animals and affect our lungs. Especially at risk are people who suffered from asthma or bronchial diseases and chest infections. Use your car less and reduce your own use of chemicals, and you will help to eliminate smog.

Spread the word about pesticide poisoning

Doctors know very little about the treatment and diagnosis of pesticide-related illnesses, yet as many as 250,000 people worldwide may be affected. Farm workers are exposed to a potentially dangerous chemical cocktail of poisons used in our intensive agricultural system. Cases of mass poisoning have been reported all over the world. The public is at risk through ingestion of contaminated foodstuffs, especially unwashed fruit and vegetables. They are even at risk when walking in the countryside. The effects of pesticides range from cancers and diseases of the central nervous system to kidney failure, skin diseases, and even memory loss. There is only a handful of specialist hospitals that understand the full effects of chemical poisoning by pesticides. Their possible effect on the next generation is not clear at all. Demand more research.

Reject the vitamin supplement

Herbal extracts, vitamin supplements, mineral tablets, and other health supplements are a healthy business, but they may not have any effect on us. If you eat a generally good diet, you shouldn't need any of them. Most good diets produce more than enough of the minerals and vitamins that our bodies need. Only those who smoke cigarettes, don't eat properly, and take other tablets are likely to need multivitamin tablets. Mineral supplements are unnecessary. We need only a few millionths of a gram of most elements, and they may be toxic to our bodies if taken in tablet form. Deficiencies are almost unknown in industrialized countries. Don't bother with supplements: check your diet.

Slow down naturally

The hectic and fast-moving life-style of the 1990s will bring with it problems that will take more than drugs to solve. Millions of people are said to be addicted to tranquilizers to help them cope with the stresses and strains of their unnatural environment. The Department of Health has urged that all tranquilizers should be prescribed for shorter periods and for fewer conditions. Alternative and holistic methods for calming your body down include relaxing massages, deep breathing and meditation, herbal teas, drinking less coffee, giving up additives and sugar, and finding counselors with whom to talk through your problems.

Walk it

Walking briskly 1 mile every day can make you fitter and you could lose one pound of weight every week. No pollution, no fuss.

Protect the rain forests

There are about 3,000 plants with known cancer-curing properties, and about three-quarters of these grow in tropical rain forests. There may be many more. We may never know the true cost of the destruction of these precious ecosystems.

Say no to drugs

Doctors often give us medication in the form of drugs because they think we expect it, not because they know it will cure us. Cut down on the number of chemicals you accept into your body, and reduce the production of potentially harmful drugs.

Cut down on fluoride

Fluoride was considered a poison until Italian dentists in the early 1800s found that it helped to reduce cavities in the teeth, although its use often led to unsightly teeth. In the 1930s metal-refining industries found that they had to dispose of quantities of corrosive and dangerous fluoride, and they came up with the idea of selling it to water companies. The body can tolerate tiny amounts of fluoride without serious side effects, although more than 4 milligrams per day can cause brittle teeth and abnormal bone formation. Reject fluoride in toothpaste, and campaign against it in water. You get quite enough fluoride from tea, because of the soil it grows in, and from fresh vegetables.

Cut down on chlorine

Chlorine and its compounds occur in bleaches and disinfectants, household cleaners, and fire extinguishers. It is used in processing wood pulp, in refining sugar, in manufacturing dyes, drugs, and oils and in the preparation of frozen vegetables. Chlorine is extremely reactive. It links up with other chemicals and can form compounds that are extremely dangerous for the environment and for your health. Many people are sensitive to chlorine without realizing—notice your reaction in swimming pools, for instance? Cut out chlorinated bleaches and refined foodstuffs that use it, and campaign for chlorine-free paper.

Reduce your solvent intake

Many products contain solvents: polystyrene and plastic products, decaffeinated coffee, foods containing glycerol (such as ice creams, cakes and cookies), ink, lipstick, dyes, detergents, and deodorants. Some are more toxic than others, but together they represent a considerable daily exposure, and their continued manufacture poisons our environment. Look for nontoxic alternatives.

Breathe!

Most people don't breathe properly. We use only part of our lungs to keep us alive. Breathing properly will massage every organ in your body. This will do more for your health and well-being than a handful of assorted tablets and pills, and it's free. Take time off everyday, practice breathing at bus stops, on the train, while you wash—very soon it will be second nature.

Check your dentist

Some people who have had fillings removed find that they feel more alert and think more clearly and that their memory improves. The fillings contain mercury, silver, copper, tin, and sometimes zinc. Chewing releases small amounts of filling into the food in your mouth and over a long period of time may poison you. In 1987 the Swedish government banned the use of mercury fillings in the teeth of pregnant women because of the risks to the fetus. It saw the ban as the first step in its program to phase out the use of mercury in dentistry. Mercury fillings in teeth can affect the immune system. Ask your dentist to use an alternative compound.

Laugh

Don't take the world too seriously. Laughter can help lighten the load—laughing relaxes your whole body, deepens breathing, expands blood vessels, improves circulation, and speeds tissue healing. We may not be able to solve all our problems with laughter, but we can certainly make the world a more pleasant place to live.

Talk about it

Don't bottle up your misery. Talking your problems through with a friend, a counselor or a member of the family could be a lifesaver. A good listener will help to solve some of your worries immediately and will guide you on the road to healthy confidence and self-esteem.

Death

Start a "Trees for Remembrance" scheme

When loved ones die they can leave behind a future for tomorrow. Plant trees as life and death celebrations that will serve too as living monuments that everyone can enjoy.

Ash is no fertilizer

Scattering the cremated remains of a body around a rose bush or flowers won't help them to grow. It might even kill them. Disperse the ash over as wide an area as possible, or ask the crematorium to dispose of the remains for you.

Choose sustainable coffins

At least 80 percent of the 600,000 coffins used in Britain every year are made from chipboard and veneer, imported from Belgium. Most of the veneer comes from tropical rain forest sources, and the chipboard itself is bonded with pollutant formaldehyde glues and resins. The remaining 20 percent of coffins are made of oak from North America. To make an environmentally sound choice we should stop importing tropical hardwood veneers, such as the 2,600 cubic yards (2,000 cubic meters) that came from Belgium during 1988, and campaign against the use of formaldehyde, which is a suspected human carcinogen and irritant. Buy North American oak if you can.

Avoid PVC

Polyvinyl chloride (PVC) wraps for bodies inside coffins are causing an environmental problem. Concern has been expressed about the high levels of dioxins and polychlorinated biphenyls (PCBs) measured around crematoria. It seems likely that the PVC wraps render the emissions even more pollutant.

YOUNG GREENS

Children and
Young People

Use your head

Children can quite easily pick up nits or lice from other children at school. If your child has become infested, the first thing you're likely to do is to rush down to the drug store. But the chemical mixtures in head-lice shampoos include lindane, carbaryl, and malathion, which are pesticides. Lindane is the most dangerous and was banned as an ingredient in sheep-dip some years ago. Try getting rid of the lice using nontoxic alternatives such as cider or malt vinegar, or the oils of rosemary, eucalyptus, geranium, lavender, or coconut. Rub the oil into your child's scalp and then use a steel comb to remove the dead lice. Applying natural oils is not only much safer but a cheaper remedy, too.

Don't let young children play near busy roads and motorways

Children like the sense of thrill and adventure associated with fast cars but, playing by the roadside, they are often oblivious to the very real dangers involved. Quite apart from the obvious risk of an extremely serious or even fatal accident, there are also some invisible threats. Brain disease, asthma, colic, palsy, and anemia are illnesses that may be triggered by exposure to the fumes of leaded gasoline and other exhaust gases emitted by cars.

Reject disposable diapers

Americans use 18 billion disposable diapers each year. But this
convenient product has enormous environmental implications.
For a start, one average-sized tree must be felled to produce
between 500 and 1,000 diapers. The paper pulp is chlorine-
bleached, causing problems with dioxin contamination. The
use of a mixture of plastics and synthetic absorbancy chemicals
is still part of the production process. In short, the cost—
economic and environmental—of using and disposing of these
diapers is enormous: they make up roughly 4 percent of our
household waste; and studies from the United States have
shown that they take over 500 years to biodegrade. So, save
yourself money and help the environment. Use recyclable cloth
diapers instead whenever you can.

Use a diaper service

Diaper services are on the increase in North America and have
many advantages. They are good news for your baby, who gets
clean, comfortable diapers; less energy is expended as the diapers are washed collectively; and because the diapers are non-
disposable, this means no plastic and chemicals and saves trees.
Using a diaper service also works out cheaper than stocking up
on disposables. What's more, you no longer have to wash
diapers yourself: they get collected each week and swopped for
bright, clean sterilized ones. Contact the National Association
of Diaper Services for the nearest location.

Give greener rewards

It is difficult enough being a parent, but knowing whether to
give sweets as a reward for good behavior to small children can
be a real dilemma. Sugar aggravates tooth decay, and giving
your children sweets, chocolate bars, fizzy drinks, and cookies
is, as far as the body is concerned, a punishment, not a reward.
So, isn't the risk to your child's health, on the one hand, and
the cost of unnecessary, expensive packaging on the other,

rather a high price to pay for the few minutes of relief you get from the pacified child chewing on a sweet reward?

Serve them "real" snacks

Why not serve your children "real" food from an early age? Cut up fruit and vegetables—oranges, apples, carrots, etc.—into bite-size chunks for snacks for young children and offer them raisins and nuts when they get older.

Wean them off baby foods

Baby-food labels marked "sugar-free" or "low sugar" can still contain up to 25 percent pure sugar. Government recommendations that the amount of sugar in baby foods should be reduced have only encouraged manufacturers to seek out alternatives, which they do not disclose, such as glucose, artificial sweeteners and fructose. Baby food should be entirely sugar-free and, until companies agree to use correct labeling, you should avoid them. If you prepare babies' food from the natural foodstuffs you eat yourself, their diet will be far healthier *and* cheaper.

Breast is very much the best

Breast-feeding is the best option for the environmentally conscious mother, as well as the safest for the baby. It is estimated that 3 million mothers bottle-feed their babies, which results in over 70,000 tons of waste from discarded packaging. Companies selling bottle formula for feeding babies can't reproduce the exact ingredients of breast milk, even with additives. Yet despite the benefits, in Western countries the biggest disincentive to breast-feeding is having to contend with the embarrassed reactions or downright insensitivity of many people, including doctors and health officials. If you want to breast-feed and are at all concerned about it, contact support groups before you have your baby.

Avoid using a personal stereo

Early in 1990 the National Deaf Children's Society tested all the personal stereos on the market in Britain and found that every single one of them were capable of an output of 90 decibels, the equivalent to the noise of a pneumatic drill. This is particularly dangerous for children's ears as sounds entering are condensed into a much smaller and more delicate ear canal. Furthermore, the sounds are refined to a much higher pitch, so the level of pressure in the ear is greater. A vicious circle then ensues: listening to too loud music impairs our ability to detect higher frequencies (the ones we use to ''listen'' to music with), so we have to turn the noise up in order to hear it. Apart from being potentially dangerous to the individuals using them, personal stereos also cause noise ''pollution'' for other people. All told, this alienating ''toy'' can be more of a hazard than it seems.

Buy them green toys

Environmentally conscious consumers are beginning to turn away from the gadget-type toy to more traditional, well-built items like wooden dolls and teddy bears. This trend implies an enormous saving of resources in the long run, with a reduction in the numbers of broken, gimmicky toys, their batteries all run-down, that children discard so quickly after Christmas and birthdays. By contrast, well-made teddy bears can last for generations and indeed have recently risen to the status of collector's items.

Get them to make things

Instead of mass-produced toys, why not encourage your child to use a bit of imagination? They can make things from boxes, empty containers, scraps of cloth, beads, and natural things. You'll save money and lose your sense of dismay at seeing children taking expensive toys apart. This approach can also help to cut down on some of the wasteful packaging of young children's toys.

Hair is good for the garden

Use hair clippings on your garden. The nitrogen in hair can release itself over many years, feeding trees or shrubs.

Try prenatal care

Looking after yourself before you have a child is as important as what happens to your body during pregnancy. A healthy baby, conceived in a healthy body, has a higher chance of survival and of better health in its lifetime. The water we drink, the food that we eat, and the air that we breathe all have an effect on our bodies. If you think you want to start a family, get some advice about prenatal care.

Give your child a massage

Children need and love to be touched. A soothing, relaxing massage will help to calm them down at night and get a good night's sleep. I prefer to use almond oil with tiny drops of different essential oils for different moods. Massage helps children who are anxious or depressed and can even calm hyperactive children. It also encourages bonding between parents and children. Tell them that you love them.

Avoid hazardous chemicals

Pregnancy brings with it some risks and changes in your life. Many pregnant mothers cut down on tobacco and alcohol or other drugs, but some hidden chemicals may have acute effects on pregnant women. In the United States doctors advise women against using cleaning products, hair dyes, food additives and perfumes, even tea and coffee. (The aware green person would do well to avoid these anyway.)

Cut down on ultrasound scans

The World Health Organization has questioned the use of scanning by high-frequency sound waves to monitor a fetus in the womb. There seem to be no long-term safety studies on this technology, but reports from the United States have claimed a link between scanning and hearing disorders; some have linked scans with dyslexia. Many mothers have said that their children appeared to be affected by the scan and tried to move away in the womb. Until we know more about the long-term implications of scans, a pregnant mother might question their use if they aren't vital.

Don't have X rays if you are pregnant

X rays and pregnancy don't mix. In the last thirty years we've realized that routine X rays for pregnant mothers have been responsible for cases of leukemia and other long-term cancers in children. It is now known that any exposure to radiation from an X ray can have an effect on the fetus, and children of any age should avoid X rays except in emergencies.

Avoid "disposable" babies

Millions of packets of disposable cloths, face and bottom wipes, and cleaners are used on newborn babies each week. Manufacturers promote their advantages: they are quick and easy to use and they are disposable. They do contain fewer chemicals than throwaway wipes for adults, but they are made from products like wood pulp and are pollutants.

Let your baby go bare!

Babies' bottoms are covered with plastics, disposable diapers and all manner of chemicals and creams. Diaper rash was virtually unheard of until plastic pants started to be used. Knitted wool covers offer some protection for babies when treated with lanolin and also allow the bottom to breathe, but the best idea is to let your baby go bare as often as possible, for as long as possible.

Don't let your child smoke passively

Passive smokers (those of us who do not actually smoke cigarettes but have to breathe in other people's smoke) are likely to suffer from illnesses such as catarrh, coughs, asthma, and colds. Cadmium and lead from cigarettes can be absorbed by children whose parents smoke, and reports suggest that children are more irritable when they inhale nicotine. We know that a fetus is affected if its mother smokes during pregnancy. Why don't we accept that cigarettes have some effect on small children? For your children's sake, don't smoke.

Beware what you feed your child

Food is fun, but not the synthetic kind. Desserts, sweets, jellies, and creams made with preservatives, colorings, emulsifiers, and sugars are mostly junk. They can aggravate allergies, cause hyperactivity, and increase the likelihood of other illnesses in children. Reject them.

Let them grow

Children love to grow things. It helps them to learn about nature and the world they live in. Encourage them to grow seeds, pips, and bulbs at home. Grow fast-shooting sprouts and encourage them to eat them. They will quickly appreciate the link between what we eat and the soil.

Watch for intolerance to chemicals

A growing number of children are finding that they are allergic to milk and milk-based products, sugars, and synthetic foods. Some doctors have been slow to recognize the symptoms; you, the parent, will be the best judge, so take note of what's wrong with your child, and take action where necessary.

Recycle clothes

Is it really worth spending huge amounts of money on clothes for children and small babies that will be worn for only a few months? Many thrift shops and charities sell practically brand-new baby clothes at a fraction of the cost. Recycle baby clothes, and teach your child to think green at the youngest possible age.

Let them read green

Hundreds of books for children illustrate the wonders of nature, explain pollution, and explore ecology. Children's storybooks should be written in nonsexist, nonracist language, helping to develop their understanding of the world in a genuinely balanced way. If you are buying books for children, look for those that espouse good ecological and social principles.

Buy nontoxic toys

Toys with nontoxic ingredients, including paints and varnishes, are now available. Although lead has been removed from some paints, it and other harmful chemicals may be hidden in painted products. Don't buy your children toys unless you know that they are safe. Look for durable, hand-crafted toys. Don't buy those from rain forest hardwood sources, and reduce plastics, batteries, and metals to a minimum. Teach your children to reject wasteful consumerism at an early age.

Don't let children sit too near the television

Nonionizing and electromagnetic radiation from television sets is still under scrutiny. The only thing that scientists agree on is that children shouldn't sit right in front of the television, as the radiation seems to be strongest there, though it quickly dissipates after 6 feet (2 meters). In Sweden manufacturers have come up with a model that cuts the radiation by half. Lobby for low-radiation models worldwide.

School and College

Raise funds for good environmental projects

Sponsoring positive environmental schemes abroad and at home is an excellent way for students and pupils to raise money and learn in the process. There are thousands of tree protection and planting schemes, seed schools, water-purifying projects and small solar schemes to promote.

Twin with an ecology-conscious school abroad

By twinning with a similarly ecology-conscious school or college you could swap notes on local environmental problems and successes and educate yourself about the world at the same time.

Plan a nature garden

You can learn so much about nature and wildlife from just a small patch of land: select plants for size and color, learn about seasons, extend original habitats and create new ones for birds and butterflies, and grow herbs for food. A worthwhile venture will need to have continuous involvement of teachers and parents and rosters for maintenance and watering.

Adopt a tree for lunchtime

Schools in Australia have been pioneering a new lunchtime exercise. Children "adopt" a tree, and throughout their school or college life they eat under it, rest under it, and care for it.

Go for violence-free science

During 1986 over 750,000 animals died in experiments in schools, universities, and polytechnics. Students and campaigners are urging a ban on this cruel method of teaching. They are encouraging the study of science without animal abuse, and they are campaigning against any disciplinary or administrative action that could constrain students who wish to reject experimentation on animals.

Campaign for vegetarian school meals

Get your school to provide a range of alternative vegetarian meals at lunchtime. Lobby your local authority; get up a petition; get teachers on your side; ask your parents to write letters on your behalf; and send in menus that exclude meat dishes.

Ban additives from schools

A study by New York City's state schools during the first part of the 1980s concluded that children who eat fewer additives and chemicals are brighter and less aggressive. Over 1 million children took part in the study, and it was found that the average increase in academic ranking was 15 percent among children who were served meals that were healthy and additive-free. For some children the only decent meal each day is the school lunch; local education authorities should be especially vigilant that tomorrow's generation has the best possible food.

Write and paint

Education about the environment isn't gained from textbooks alone. Share your concern about the plight of our planet by expressing your views in essays, stories, and poems or by painting and drawing. Use your school magazine as a vehicle for disseminating information and stirring consciousness.

Negotiate a noise agreement

Noise is a polluter; it should be recognized as an environmental problem. Left unchecked, it can result in deafness in extreme cases. Caring for our bodies must include caring for our ears, and learning about the pollution from noise is an important beginning. Negotiate a noise agreement with your parents, school, and neighbors, and keep everyone happy.

Test the pollution levels

Science labs are good places to start learning about pollution, and the environment around your school building is the place to begin. Test for acid rain, for instance, or do a survey on the amount of litter in the local area. Work out how much it would cost, in terms of money and energy, to put things right.

Campaign for better public transportation

Most young people at school or college travel by public transportation. Walking used to be fairly safe, but more and more accidents are taking place as roads are widened and schools are amalgamated to save money. Campaign for public transportation that is safe and more friendly to the environment.

Use solar-powered calculators

If you use a calculator, reject batteries, which use fifty times more energy to produce than they actually provide—use a solar-powered calculator instead. They are the same price and never need renewing. They work even without direct sunlight; light from bulbs is enough to power them.

Make fast food a no-no

Fast food in the form of hamburgers, French fries, and hot dogs should be unacceptable to young people, who deserve a good diet. Peer pressure is the fastest way to stop the rot; green students will know of the problems associated with fast foods, which contain preservatives, colors, artificial ingredients, and large amounts of fat and sugar—and are covered with wasteful packaging. Tell the others.

Plant a tree for each new class

Each year a new class of students arrives at every school and college. What better way to celebrate their arrival and be positive about the environment than to plant a tree for them? Perhaps the outgoing class could plant the tree for the incoming one? If there isn't space in the school grounds, ask the local authority or parks department to allow you to have a small piece of land.

Learn about Third World aid and development

The predicament of the so-called developing countries is a cause for concern around the world. Students should know that the life-style of the First World is responsible for chronic problems of these countries—not to make them feel guilty but to empower them with information and to change their attitudes, so that tomorrow's generation will understand that we all share the natural resources of this planet and are responsible for husbanding them.

Bring back the fountain pen

Throwaway pens have become the norm in schools, colleges, and universities. We throw away literally millions of them. While no one wants to go back to the chalk-and-slate era, refillable ink pens will cost less and use up fewer resources.

Invite speakers to share their knowledge

Why not invite speakers from environmental or animal groups to tell you their story? They will help to widen the environmental debate and will supply information. Welcome speakers from industry and government, too.

Make schools and colleges no-smoking zones

Most smoking starts in schools. Pressure from friends encourages many young people to start smoking and get hooked for no reason other than to keep up with their friends, who probably don't like it much either. The biggest contribution you can make to your health at school is to persuade everyone that smoking is socially unacceptable.

Attend cycling proficiency classes

Avoid accidents on the road; learn to cycle properly. After all, a car driver has to pass a test to get a license. The least you can do is learn to ride a bicycle safely and well.

Check your computer

Now that many schools and colleges use computers for mathematics, computer studies, and even languages, the issue of low-level radiation from computers is an important one. Sensible chairs, adequate ventilation, and appropriate lighting are crucial; VDTs should be checked for radiation emission; and exposure should last for no more than an hour at a time. Good working practices should start at school, so that they carry on into adult working life.

Collect rubbish and save money

The collection of cans and scrap teaches young people to be aware of recycling and the dangers of litter but can also help obtain much needed income for school facilities. Negotiate a good price for the goods you collect. If people are prepared to offer decent clothes or bric-a-brac for collection, hold a rummage sale.

Put up a list

Pin up a highly visible list of environmental groups and charities, local and national, including local recycling centers and facilities, so that people will be able to contact their favorite groups.

Organize quizzes and competitions

Quizzes and competitions sharpen the brain. Why not hold an environmental quiz and test the knowledge of the greenie smart alecks?

Campaign for the return of the washable cloth

In washrooms and toilets across the country washable cotton cloths are being replaced by paper towels. These are energy intensive: campaign for the return of the cotton cloth. Campaign for toilet paper that is recycled and chlorine-free.

Organize environmental days out

Days out are few and far between at schools, but they represent an important part of the year's curriculum: pupils learn more, faster, when a day out is involved. Whatever the reason for the next day out, why not extend the experience by asking to visit a local nature park, a windmill, or an energy-saving center?

Write to manufacturers

You will have to learn how to write cogent letters anyway, so why not start flexing your consumer muscles by writing to manufacturers and companies that produce environmentally unsafe products, asking them to change their ways? Learn how to argue persuasively by writing letters of complaint.

Check the candy store

The candy store may be the provider of many over-sugared, additive-laden sweets and snacks that are destroying your teeth as well as pouring chemicals and waste into your system. Ask for quality snacks, fresh fruit, mixed nuts, and raisins to be added to the candy store list. What about carob as an alternative to chocolate, or freshly squeezed orange juice instead of the sugared variety? What about sugar-free sweets?

Ban aerosols

Ban aerosols from your school or college: declare it an ozone-friendly area. Don't forget that substitutes for CFCs in aerosols are also either ozone-depleting or hydrocarbons, contributing to the greenhouse effect. The safest environmental option is to ban them all.

Have save-a-scrap days

Collect scrap materials such as egg boxes, unusual clothes, containers for soap liquids, and the like for primary and pre-school groups, which love making things with them. Scrap can be useful when it is transformed into material for make-believe.

Sign petitions

Get involved. Collect and sign petitions on environmental issues, and voice your protest with others.

Grow plants and flowers

Don't just grow one plant in a bottle: be adventurous and make your school or college grounds something to be proud of. Plant native flowers and vegetables. You can sell them and cuttings at fairs or rummage sales to raise cash.

Make videos of local wildlife

Many schools and colleges offer access to video or photography equipment. If you have the chance to use it, you could present a valuable record of local wildlife for generations to come as well as having fun learning.

Debate the issues

If your school or college hasn't already got one, start a debating society. A debating society will help you learn to articulate your concerns about this planet.

Do an environmental-impact study on your school or college

You could learn how to assess environmental impact by starting with the building in which you learn. An environmental-impact study could look at the waste you create, the amount of energy that is consumed there and even the effect of the equipment you use both on the local environment and from the moment that it is created until it is disposed of.

Form a group

Get together with a few friends and start an environment group. Discuss possible actions, meetings, and campaigns, and teach each other. Join up with organizations that run youth groups, like Friends of the Earth.

Use only recycle paper

Campaign to get your school or college to use only recycled paper in the building. Some school notebooks are already made from recycled paper, but the number could be increased dramatically.

Collect paper

Collect notebooks from students when they have finished with them, and add them to a paper collection in the school or college grounds.

Get environmental studies onto the core curriculum

Many schools and colleges now offer environmental studies but not as a core subject, which means that the subject is not compulsory. Ask for environmental studies as a core subject, so that proper exams will heighten young people's awareness of the planet.

Campaign for a green library section

Your college or school could set aside a special green section that contains books, magazines, and newspapers about the environment.

Experiment with care

Chemistry laboratories should have a policy of recycling, or safely disposing of, all chemicals instead of pouring them indiscriminately down the drain to add to water pollution. There are regulations that govern the disposal of chemicals in schools. Are your science teachers observing them?

THE GREEN
CONSUMER

Shopping

Reject "biodegradable" plastic

The plastics industry has begun to fight environmental criticisms of its products by introducing so-called "biodegradable" plastics. These new starch-based plastics are expected to degrade into carbon dioxide and water when buried in wet soil. The green shopper should not be conned by these marketing promises as some of the "biodegradable" plastics are only broken down into pieces so tiny that the human eye can no longer see them; these are called "biodestructable" plastics. To replace small plastic polymers with starch ones literally means taking food from the hungry to appease your conscience. It is also unlikely that these plastics will degrade as quickly in landfill sites. Plastics should be recycled and not left in garbage dumps to biodegrade. Biodegradable plastic takes more energy to produce, it is not as durable as ordinary plastic and has less chance of being reused.

Consider if you really need to go shopping

Shopping is said to be one of our favorite pastimes. During the consumer boom in the 1980s we heard of the compulsive shopper, the professional shopper and the functional shopper. In the United States $3,191,353,425 is spent every day on shopping, and a proportion of that is for food and clothing that will never

121

be eaten or worn. We waste food, chemicals, and money spent on transportation and raw materials in this way. The next time you feel like shopping, ask yourself if you really need to go. If you do want to go window shopping try leaving your checkbook, money, and credit cards at home.

Buy in bulk

Buying goods in bulk does not only save you money, it also saves on packaging, transportation and storage, and it cuts down on trips to the shops. Try finding a good cooperative that buys vegetables and grains in bulk or buy fresh food when you visit the countryside.

Read the label

You should get into the habit of reading the labels on your foodstuffs. Check for added salt and sugar which you may want to avoid. All canned fruits and vegetables I inspected in one major store contained sugar. Look for food that has been produced without artificial chemicals or additives. Find out about the country of origin; you may want to boycott the produce of particular countries.

Buy simple

With so many thousands of products to choose from on our supermarket shelves, the average shopper can be forgiven for being confused and buying too much. When shopping, you will have to come to some decisions about the environmental problems facing you and buy only what you need. To help cut down on pollution, buy products that have been processed as little as possible and have traveled the least distance.

Get out of the throwaway trap

You know that when you buy good quality, durable products they may cost more but will save you money as time goes on. It also helps the environment in many different ways by cutting down on the raw materials, packaging and processing of cheap mass-market goods, reducing pollution and billions of tons of waste. Always complain to manufacturers and shops who do not provide good, durable products.

Go without the carton

Cartons of milk and soft drinks may save energy in transportation as they can be stacked to take less room than, say, glass bottles. But the carton is not reuseable, recyclable or biodegradable. It is made from layers of highly bleached paper, aluminum, and plastic which poison and pollute the environment from their processing until the end of the carton's non-reuseable life. We should be pressing for a better environmental alternative such as the reuseable glass bottle.

Shop at cooperatives

Shop within a cooperative and you will be supporting good businesses and small-scale manufacturing processes that are good for the planet. One of the most famous consumer cooperatives is the Japanese Seikatsu Club. It has 300,000 member households and is run and organized predominantly by women. The club sells household products without adding profit and the fair price paid by the buyer is passed on to the farmer or producer. The cooperative does not need mass advertising but informs its members through a regular newsletter. Only one product in each line that meets strict environmental and health criteria is chosen. Cooperatives offer products based on sharing and need rather than profit and greed.

Buy recyclable packaging

The way your food is packaged is crucial to environmental considerations. Paper packaging that also contains strips of aluminum and plastic and Styrofoam containers are impossible to recycle. Simple packaging of single layers of paper, steel, aluminum, glass and rubber is recyclable, but it is not worth buying them if you simply throw it in the trash can afterward.

Do not buy goods packaged in more than two layers

There are virtually no products on the market that need more than two layers of packaging. Take a box of chocolates, for example. It can contain more packaging than actual sweets. Beauty products, gifts, games, and sweets are the main over-packaged products. Most of the money you pay goes into the extra layers of packaging.

Refuse the plastic carrying bag

Next time you go shopping, take your own carrying bags or tough and durable shopping bags. Billions of plastic bags, none of which are recyclable or biodegradable, are given free to shoppers everywhere. There is a simple, green alternative: refuse and reuse.

Don't buy polystyrene packaging

Reject polystyrene packaging which contains chlorofluorocarbons (CFCs) and damages the ozone layer. If the package is not labeled, then assume it does contain CFCs. Even the new packaging contains HCFC-22, or hydrocarbons that are called "soft" greenhouse gases. Buy loose food, take your own bags, or use recycled brown paper bags.

Write letters

The pen is mightier than the sword, they say, so why not use it? Write to packaging manufacturers and ask them to forgo CFCs and excess packaging on items you buy regularly. Write to manufacturers of canned goods and ask them to remove the sugar. Write to paper suppliers and ask them to produce un-bleached and recycled paper. Manufacturers will have to take notice as you are the customer and they will not have a market if you do not buy their goods. Get information and statistics to back up your request; always be polite and ask for a reply.

Don't be fooled by the freebie

Gas stations give us toasters and fancy glasses. Cosmetic companies offer expensive towels and accessories. High profit margins and fierce competition have made it worth their while, but many companies are now beginning to regret the trend, as customers seem to be more interested in the free gifts than in the products themselves. Don't be fooled by the free gift; ask yourself if you really need the product offered.

Buy from shops that have environmental policies

As consumers become more aware, many supermarkets and shops are now considering environmental policies. Choose your goods from shops that have stated policies for their own-label produce and that also require high standards from others. Honest and informative advertising and labeling are the fastest ways of ensuring you are well informed enough to make a choice. Ask for these; they are your right as a customer.

Support the small shop

Out-of-town supermarkets and shopping malls with huge parking lots have pushed the small, local shop to virtual bankruptcy. Small shops save on energy: you don't have to drive, you just walk there with a shopping bag. The shops are more likely to offer local produce and they provide a vital lifeline to people without cars.

Buy a rechargeable battery

If you must have a battery-powered product, then you should buy rechargeable batteries whenever possible. You can recycle them up to 500 times using a cheap battery charger. Although they save money and resources, rechargeables do contain cadmium, a highly toxic material, which is gradually being phased out of battery manufacture in Europe. One-third of the planet's cadmium goes into batteries, and the environmental effects are extremely serious. It is likely that manufacturers will come up with a safer rechargeable battery in the future, and we should encourage them to do so.

Ask for labels on fresh foods

Labeling has become important to today's shopper, but mostly for prepackaged foodstuffs. Ask for labels to tell you about the pesticides that have been sprayed on unpackaged foods and also for information on how to wash off or get rid of unwanted residues. This action could spur more producers into dropping their use of some pesticides and you, the consumer, will be better informed.

Shop for an end to oppressive regimes

We often feel that we can't do very much when we see on television and in our newspapers pictures representing some of the barbaric and oppressive regimes that operate throughout the world. The human suffering caused by the apartheid regime in South Africa has made many of us aware of racism and of our own political strength when it comes to boycotts. When you shop, you can show that you care by not buying products from countries with oppressive regimes.

Go to rummage sales

Pick up cheap secondhand goods. It's a form of recycling!

Buy a rain forest

Well, not really. You can support some rain forest groups that buy "protection" for rain forests. They should not be anyone's to own, but protecting the rain forests means that indigenous people can carry on living there and that the area will not be developed by industry or cattle grazers. Organizations like the World Wide Fund for Nature operate such schemes.

Beware the green label rip-off

Surveys by consumer groups have pinpointed a new rip-off in the marketing world: the environmentally friendly product. Companies make excessive claims about products like aerosol cans that are "green" and batteries that are "environmentally friendly" because they have withdrawn the mercury and cadmium content. My view is to avoid products with the "environmentally friendly" label. It usually means that there is one good thing about them, but potentially many other things are wrong. Companies that produce truly "friendly" goods do not need to say so quite as loudly: their ingredients label and the information on the packet tell you everything.

Swap with friends

Swap clothes, goods you no longer need and other products with friends. Reinstate the art of reusing clothes and goods, save yourself money and have fun in the process.

Don't be fooled by advertising

Advertising products is big business. A company that can afford to advertise on television is usually promoting a highly manufactured and manipulated product. The cost is always passed on to you, the customer. Watch the advertisements to judge the product for yourself. In North America 47 percent of viewers find television ads annoying and unnecessary.

Look out for planning controls

Supermarkets are springing up everywhere. The land used is often just outside a town and large areas are concreted over. Before planning permission is granted the plans have to be assessed, and at this stage you can have an effect on the building. You may want to oppose it altogether or you may just want to get the developers to agree to certain conditions: for example, to keep the trees on the site.

Campaign for play areas

Taking children shopping can be a hassle if they stop you from getting on with what you want to do. At supermarket checkouts sweets, chocolates, cookies and chips are stacked low on shelves to entice children. Play areas could save you both stress and money. Campaign for a play area in larger supermarkets and get smaller shops to join together to provide them.

Give back the extra packaging

In Sweden toothpaste boxes will be phased out completely by the end of 1992 as this extra packaging is unnecessary and wastes energy and resources. When you next visit the supermarket, try leaving unnecessary packaging behind, or at least write to the manufacturers to complain about it. Someone will soon get the message.

Support the reusable

Support companies that produce products and foods in reusable containers like milk bottles, refillable plastic containers at the Body Shop, returnable soft-drink bottles and packaging with a deposit to encourage you to bring it back.

Eating

Eat out at health food or vegetarian restaurants

Many restaurants are beginning to recognize the advantages of serving vegetarian and health-food meals. My most rewarding meal in the United States was a visit to the legendary Moosewood Restaurant in Ithaca, New York, a vegetarian restaurant that cannot be beaten. There are plenty of guides to other good wholefood restaurants around the world.

Cut down on chocolate

Britain leads the world in chocolate consumption. The number-one-selling sweet product is Kit Kat, followed closely by the Mars bar. Between them they generate sales worth £300 million a year. Americans eat 1,000,000 tons of chocolate each year. Cocoa, the main ingredient of chocolate, has come under the scrutiny of environmentalists recently because of the highly dangerous pesticides used on the cocoa bean. The chocolate bar itself is highly processed and sugared and this affects teeth, digestion, and general health. Chocolate was favored by the Quakers in the early eighteenth century as a healthy alternative to alcohol, but today's processed, pesticide-laden variety is hardly the same thing. Drink the purest cocoa you can find; eat only chocolates sweetened with fructose (from pure fruits); and generally buy less chocolate.

Eat brown bread

As early as 1826 doctors warned of the health risks associated
with a diet based on white bread. White bread is highly processed and robbed of much nutritional value. Fifty-nine percent
of bread eaten in the United States is white. In the early eighteenth century only the wealthy ate white bread, and bakers
would add lime, chalk, and animal bones to make bread whiter
for the masses who wanted the same bread as the wealthy. The
plain reason for processing brown bread into white is not to
give you a better product, or even a wider choice, but to cut the
cost of baking and to make higher profits using lots of air and
water. Eat organic brown bread whenever you can.

Eat real food

By buying preprocessed foods that are convenient and quick
you make yourself unhealthy as well as allowing the food-
processing industry to keep millions of the poor in developing
countries hungry. Lax controls, poor research, questionable
ethics, and the addition of chemicals and ingredients that are
known poisons are a real scandal. Most of us do not think about
the food we are eating. We take it for granted that someone
"out there" will have our interests at heart (they haven't always). But, as consumers, we have become addicted to con-
venience food. We expect cheap, attractive-looking food,
packaged and clean, so food preparation does not interrupt our
lives. We could save so much energy and avoid pollution if we
started to choose food with a more discerning eye. Over 75
percent of the food North Americans eat is processed in some
way and in energy terms, this costs over $10 billion each year.
Food processing adds directly to pollution, acid rain, the greenhouse effect, and a host of other environmental problems.

Buy fruits and vegetables in season

Buy fruits and vegetables in season and save energy on transportation, which in turn cuts down on pollution. We expect
fruits such as bananas and oranges all year round. Many of the

more exotic fruits are bought because they are on the shelf in front of us, transported thousands of miles out of season for us to eat. They are likely to have been treated with pesticides and chemicals to keep them fresher for longer. If you don't buy fruits and vegetables until they are in season, you will enjoy a better diet, too. Every item of food that is grown and consumed locally is an important addition to your diet.

Boycott Faroe Islands fish

An all-out boycott of fish from the Faroe Islands was called for by environmentalists in 1989 after unsuccessful attempts to persuade the Faroese people to stop killing the pilot whale. The ritual slaughter of thousands of whales by the Faroese is not to provide subsistence food. The reason, groups like the Whale and Dolphin Conservation Society say, is sport. The whales are dragged to the shore with large hooks and incisions are made in their blowholes, causing a lingering death. The British and United States market take 90 percent of Faroese fish.

Cut up your vegetables

If vegetables, meat, and other foods are cut up as small as possible before cooking, you will find that they take half the time to cook and save energy and money as well. Potatoes for mashing can easily be cubed, for example, and carrots can be thinly sliced.

Boycott California grapes

The United Farmworkers Union in the United States has called for a boycott of California grapes because of the enormous suffering of workers using the pesticides: captan, methyl bromide, and parathion. The 300,000 reported cases of poisoning per year do not even include the cancers and birth defects the workers and their families suffer. But the local environment suffers, too, as groundwater contamination affects the local community and wildlife. Captan, one of the fungicides used, has been banned or severely restricted by Finland, Norway, and Sweden, and tiny amounts cause harm to fish.

Do without dried potato

The overpackaged "space-age" dried potato is definitely not food for a greenie. Given its content of pyrophosphate, added vitamins, salt, dried whey, emulsifiers, preservatives and antioxidants, you could be forgiven for asking where the dried potato is! The energy you spend boiling an electric kettle to pour water on the dried version is comparable to the energy used to cook a potato on a gas stove to mash yourself, so why not buy the real thing?

Reject irradiated food

Good fresh food does not need to be irradiated with cobalt-60 or caesium-137 to give it a longer shelf life. Environmental groups claim that food irradiation is just another way of using up nuclear waste, the irradiation source. Irradiation does not kill all the bacteria that are likely to be dangerous (for example, *Clostridium botulinum* is not affected, and eating food contaminated with botulism can be fatal). Irradiation also affects foods in different ways, most notably in terms of flavor. It can reduce the flavor and vitamins in food and can destroy essential fatty acids. Some supermarkets have agreed not to sell irradiated food, so look out for them. If in doubt, ask before you buy.

Do without a beef burger

Tumble-massaged, stripped animal carcasses, gristle, sinew, heart, tongue, and fat added to colorings, flavorings and preservatives—or just a beef burger? Most beef burgers will have come from farmed animals, intensively reared, that eat up to eight times their body weight in grains each year. Doing without beef burgers saves land and resources, and will give you a healthier diet, too. Try a soybean burger or another beef-free alternative.

Buy organic bread

Some supermarkets supply at least one kind of organic bread, and you can buy locally produced loaves from your health-food shop. Over 12 billion loaves are consumed in the United States every year; it's time we stood up for our staple food. The whole-grain wheat used for making bread is damaged by pesticide residues even before anything else is added to it. My local supermarket displays a notice on the bread shelf that reads: BREAD AND FLOUR PRODUCTS MAY CONTAIN ONE OR MORE OF THE FOLLOWING LISTED ADDITIVES: ANTIOXIDANT, ARTIFICIAL SWEETENER, COLOR, FLAVOR ENHANCER, FLAVORING PRESERVATIVE OR FLOUR IMPROVER. Obviously bread is not a simple product anymore. Whole-grain products, especially bran-based ones, have high pesticide residue levels—unless they are grown organically.

Campaign for real fruit

We buy fruit from supermarkets expecting it to be like boxes and cans, perfectly manufactured and without blemish, every piece tasting and looking like the others, with the same color, texture, and weight. Though there are 7,000 varieties of apple registered in the United States, only eight dominate our shops and markets. Fruit orchards used to be abundant; they offered not only apples, but also pears, cherries, plums, damsons, and mulberries, with beautiful names like Cornish Honeypin, Beurre d'Avalon and the Scarlet Crofton. Find out about local varieties and ask your grocer to name the varieties displayed. Buy organic, and shop around for "real fruit" again.

Boycott yellowfin tuna

The yellowfin tuna congregate beneath schools of dolphins—primarily spotted and spinner dolphins but also stripped and common dolphins—in the eastern tropical Pacific. Tuna fleets find their catch by following dolphins, then hauling them

aboard or netting them in order to catch the tuna. Officially
around 120,000 dolphins die in the eastern tropical Pacific each
year, but the true figure may be nearer 250,000, severely de-
pressing several dolphin populations. By boycotting yellowfin
tuna you can say no to the companies who persist in destroying
dolphins.

Forget frogs' legs

Bullfrogs are killed in Bangladesh and Indonesia for a range of
food delicacies exported around the world. The legs are the
only valuable part of the frog and the still living torsos often try
to crawl away after their legs have been chopped off. Apart
from the obvious cruelty involved in this trade, taking huge
quantities of frogs from the wild has devastating environmental
consequences. Frogs eat waterborne pests that destroy crops. If
you don't eat them, there will be no market for those exporting
them today.

Don't eat veal

Within a few days of its birth, the calf is taken from its mother
and put in a wooden crate 5 feet (1.5 meters) long and 22
inches (56 centimeters) wide. It remains there for fourteen
weeks and is denied fibrous food. Instead, it is fed purely on a
diet of liquid milk substitute. The calf is kept intentionally
short of iron so that we can eat "pure" white meat. This
practice should be stopped immediately, and you should boycott veal completely. During 1988, 176,000 tons of veal were
produced in the United States.

Boycott abattoir-slaughtered venison

Eager to develop the venison industry, dealers are trying to get
abattoir-slaughtered venison regularized in the same manner as
cows, pigs, and sheep. The most humane way to slaughter deer
is for them to be shot on the farm by qualified marksmen.

Reject rennet

In order to turn milk into cheese, rennet, a gelling agent, is used. Most people do not realize that rennet is taken from the stomachs of slaughtered baby calves. Not all cheeses contain rennet, but you can be sure of avoiding it by buying cheese with the vegetarian symbol.

Go vegetarian

Millions of people are now vegetarian, making a stand against factory farming, intensive rearing, artificial foodstuffs, and hormones in animals. There is a large variety of foods to eat, so life as a vegetarian can be healthier and just as much fun.

Campaign to clean up slaughterhouses

Of the 1,000 slaughterhouses in Britain, 900 of them are so filthy that imports are forbidden by every country in the world. Contamination by flies and excrement (from sloppy slaughtering), and fatty machinery growing mold, proves that killing animals is a dirty business in more ways than one.

Buy real pork

Factory-farmed pigs live on concrete or slatted floors and lie in their own excrement. Breeding sows are often chained to the floor and the others are tied so tightly that they cannot take a step forward or backward. Piglets are born in a farrowing crate and the mother is only allowed to suckle for two or three weeks before the piglets are taken to battery cages to be fattened up for market. Some supermarkets have complained about these conditions, and pigs are beginning to be better treated. Ask your supermarket for information about the pork you intend to eat, or give it up altogether if you are not satisfied with their answers.

Make your own ice cream

Americans eat their way through a staggering 354,246,000 gallons of ice cream every year. Why not make your own ice cream or choose one of the natural brands now available in supermarkets?

Beware of food poisoning

In spite of modern technology, bacterial food-poisoning cases have increased by a large percentage. Poisoning was often blamed on the customer for bad food handling but most of the reported cases suggest that the food industry is responsible for dirty and unsafe practices. Hygiene and environmental standards down the whole food chain—in farms, abattoirs, factories, shops, supermarkets, and restaurants—have to be improved. An independent Health Department, separate from agricultural and food producer interests, would be a good start.

Have a great salad

Nitrates have been found in lettuce. Chlorothanonil and permethrin, probable carcinogens, have been found in tomatoes from the United States. Fifty different pesticides are used for onions and seventy-five for cucumbers, including dieldrin. The very symbol of health, the salad, has become a health hazard. Always wash salad ingredients and carefully peel everything possible. Otherwise grow your own or buy organic produce.

Reject the broiler chicken

Most broiler chickens are reared in semidarkness in huge sheds with up to 100,000 other chickens. They are genetically manipulated to produce as much edible meat as possible. At seven weeks the chickens are slaughtered by stunning them with an electrical charge in a bath and then their throats are automatically slit. Reject intensively reared chickens as there are alternatives.

Salt of the earth?

Salt has been used as a food preserver for centuries. Now with the introduction of the fridge and freezer it is used mostly for taste, and we still eat more than ten times the amount our body actually needs. Salt is added to preprocessed foods—another good reason for giving *them* up. The habit of shaking the salt over every meal is wasteful and possibly dangerous as too much salt in the body can cause heart conditions.

Eggs for whose health?

An egg scandal unleased in Britain by former Health Minister Edwina Currie in 1988 sent shockwaves throughout the industry. From an average consumption per person of 200 eggs every year, sales dropped to less than half this level and hundreds of thousands of hens had to be slaughtered. The real scandal centered around the bacteria and salmonella levels found in eggs. Hens were being fed ground up dead chickens and contaminated foodstuffs. The cruelty involved in battery-caged and intensively reared hens is encouraging more and more people to buy free-range eggs. In Switzerland a law was recently passed ensuring that all hens were to be free range: allowing them space to move around and to go outside—a luxury not afforded to battery-reared hens. Over 2 million battery hens die prematurely each year because of the conditions and the average battery hen lasts less than a year before its egg output starts to fall off and it is slaughtered. You can help directly by buying free-range eggs, cutting down on the number you buy, and by searching for the still rare organic egg.

Cut down on sugar

The high level of sugar consumption is blamed on the number of prepacked foods that contain sugar—sometimes without the consumer being aware of it. Sugar causes tooth decay and is indirectly linked to the increase in obesity, diabetes, heart disease, and even gallstones. Yet sugar is big business and even government officials do not always want to rock the boat when it comes to advising the public about health. A normal diet, especially with plenty of fruit, will give you enough fructose and sweetness without additional sugar.

You are what you eat

The average person eats pounds of chemicals and additives every year. Add to that the residues of pesticides on our fresh foods and pollution from our tap water and from the air, and consider the amount of chemicals that could have a disastrous effect on us. Many of the pesticides that were once used are now considered dangerous, but with around 1 million new chemicals being invented every year it may be many years before we realize the full danger of our industrial life-style. Food is crucial to our well-being: eating organic, additive-free food will help our bodies as well as the environment.

An apple a day may not keep the doctor away

Publicity over the use of Alar, the plant-growth regulator on apples, has prompted the withdrawal of the chemical by the manufacturer, Uniroyal. About 40 percent of our apples were said to be treated. When apples are turned into apple juice or apple sauce, a by-product chemical, UDMH, is formed which many studies in the United States have shown to be a human carcinogen. Even though the government has so far refused to ban the chemical, the recent public outcry and the manufacturer's response has ensured that Alar is no longer an unknown chemical. Other pesticides and growth regulators are used on the humble apple however, and tests have found traces of diphenylamine and captan. Captan is a possible human carcinogen and although the residues will be reduced by washing or cooking, I recommend you peel all nonorganic apples before you eat them.

Buy cardboard egg cartons

Cardboard egg cartons are generally made from recycled paper that saves energy and resources and cuts down pollution. Until fairly recently most supermarkets used foam cartons. These were ozone destroyers as CFCs were used in their manufacture. But all over the world today manufacturers are switching to less damaging materials. Use recycled cardboard whenever you can.

Nature does not create perfect fruit

When you buy fresh fruits and vegetables you generally get those chosen and approved by the big supermarket chains or markets. The produce looks perfect in every way. Often individually wrapped or boxed and without a blemish on them, you could be forgiven for thinking that they are the fruits of paradise. But this marketing is causing problems all over the world. Our standards are so high that we end up wasting food, energy, packaging, and chemicals. Do you really need every banana to be the same length, or your cucumbers to be straight? Natural vegetable molds can be easily washed off. Try to live without chemicals. It will make a big difference and will help you accept the so-called visual quality of the superior organic produce more easily.

Go without the individually wrapped snack

Every lunchtime we manage to consume millions of packets of potato chips, small cartons of fruit drinks, biscuits, and snacks. These minipacks take far more energy and money to produce than standard packs—which is why you pay more for them—and, of course, they are thus far more damaging to the environment. Take your lunch with you or buy in bulk and save the rest for another day. Buy foods that are not heavily wrapped. Cut pollution and costs, and save trees.

Go without wax

Certain fruits and vegetables have been waxed before they get
to you to keep the skin from shriveling up and to make the
product look more appetizing. If you do not eat the skin of
oranges, cucumbers, grapefruits and avocados, you might not
notice, but those of you who cook with lemon rind, for example, will be safer using the organic or unwaxed varieties. The
waxes are usually authorized to contain paraffin, synthetic resins, shellacs, and fungicides like benomyl, orthophenylphe-
nols, imazalil, and dicloran. In the United States the Federal
Food, Drug and Cosmetic Act is supposed to require all shops
to label waxed products, but this does not seem to be a priority
in many stores. Go without wax whenever you can.

Learn to love mushy tomatoes

Over a hundred pesticides can be used on tomatoes to keep
greenfly and other insects at bay. But few companies have been
able to beat the mushy tomato syndrome. Every year the average United States citizen buys about 24 pounds (11 kilograms) of tomatoes and at least 2 pounds (0.9 kilogram) are
thrown away because they are squashed. Companies are working around the clock to engineer the perfect tomato, but we
could accept battered and mushy tomatoes to make chutney and
tomato sauce.

Campaign for organic strawberries

In the early 1980s Friends of the Earth UK organized a campaign against the use of pesticides in foods. They concentrated
on strawberries and ran an ad of a blue plastic box containing
a dozen tantalizing-looking strawberries, asking: "Which contains more chemicals, the strawberries or the box?" The fears
about strawberries are real. When the Natural Resources De-
fence Council tested strawberries in the United States, they
found more pesticide residues than in any other fruit or vegetable. The residue most commonly found in fresh strawberries

is captan, a possible human carcinogen. It can be reduced by cooking and washing, but the chemical itself, already banned or restricted in Finland, Norway, and Sweden, is known to harm fish and other water life even in extremely small concentrations.

Check your cabbage

Cabbages are part of our staple diet in winter. The American Food and Drug Administration has registered over sixty types of pesticides used for cabbages, but the most commonly occurring ones are methamidophos, diethoate, fenvalerate and permethrin. Washing, removing the outer leaves, peeling, and cooking will all help to reduce these residues. Many of the pesticides are possible human carcinogens, and careful washing is vital. Whenever possible, buy organically grown cabbage.

Recycle your margarine tubs

Margarine tubs and yogurt containers are usually made from acrylonitrile butadiene styrene (ABS) or polypropylene, a type of plastic that is easily recyclable. Although the technology is available to recycle this type of plastic, for many billions of plastic margarine, coleslaw, and dessert containers the only thing we can do at the moment is to use them constructively as extra food containers, seedling and plant pots or even as children's toys.

Buy Brazil nuts

Brazil nuts need a great deal of care to grow properly. They do come from Brazil, but the farmers who have traditionally cropped them are under threat from the destroyers of the rain forests. Brazil nuts can be grown only in a rain forest setting, as they need a unique mixture of wildlife and trees in order to flourish. There is also the serious risk from aflatoxin if the nuts are not stored correctly or eaten at their best, so eat fresh Brazil nuts to save the rain forests.

Think about food

The production of livestock for our tables costs more than the life of the animal. It takes 10 pounds (4.5 kilograms) of grain to yield just 1 pound (0.4 kilogram) of intensively reared beef. It is not just a resource problem for during the Ethiopian famine we imported their cereal to feed our cattle. A diet based on soya protein could feed up to thirty times as many people as a meat-based diet. Ten acres of land will support sixty-one people on soya and only two people on beef. Nonetheless, 15 million children die each year from starvation or disease from lack of food.

Do not throw away your chopsticks

Japan, the country that uses one-tenth of all the world's wood, throws away billions of pairs of chopsticks each year. When these were invented in China and the Far East many centuries ago they were family heirlooms that were often passed down from generation to generation. One family in the forests of Malaysia complained bitterly that their lives were being torn apart for the manufacture of chopsticks while they shared just one pair, crafted by their forefathers, between a whole family. These days Chinese and Japanese food is eaten around the world; it is usually much better for you than fast food, but avoid the throwaway chopstick.

Campaign against the plastic fishnet

New fishnets made with plastics that are invisible to most birds are wreaking havoc on the marine environment. Sea birds, like the razorbill, guillemots, and shags are being hauled in the nets in their thousands. Porpoises, dolphins, turtles, and smaller fish are also caught as the trailing nets, sometimes over 2 miles long (3 kilometers), are dragged along the sea collecting every live creature possible and wiping the sea clean. Japan may be responsible for the death of over 100,000 sea mammals each year. A few countries have banned the most dangerous nets but the majority of our fish comes from nets considered unsound by conservation groups. Ask about your fish before you buy it. Get your local supermarket or fishmonger to buy only from reputable fishing industries.

Go for a natural shade of pink

Farmed fish are fed synthetic food coloring to make their flesh turn a more expensive shade of pink. The coloring is canthaxanthin. But not just trout and salmon are colored with this additive. Some preserves, pickles, fish fingers, and chicken, as well as "sun-tanning" tablets, use it. Because it is too expensive to feed farmed fish on their usual diet of shrimps, prawns, and other shellfish, they are fed instead on canthaxanthin to make them an appetizing pink.

Cut down on tin cans

Most of our cat food, canned vegetables, soup and other similar products come in a can. There are very few recycling facilities for tin in North America and Britain even though large quantities of it are home produced. Tin is not only energy consuming, but heavy to transport, and tin smelting plants have produced vast quantities of pollution. Try to buy fresh vegetables and cut down on the amount of nonessential tins you buy.

Try vegetarian pâté

Pâté de foie gras used to be the kingpin of pâtés—until we discovered how it was made, that is. It is produced exclusively from the livers of force-fed geese. Some reports say that the geese have their feet nailed to boards while they are forced to eat 5.5 pounds (2.5 kilograms) of salty fatted maize a day. Yet left to feed themselves they can produce a pâté of better quality. Leave the geese alone and try a vegetarian pâté made from mushrooms, peppers, and soya as an alternative.

Buy sea salt

Sea salt is dried in the sun and production takes half the energy it does to find rock salt from under the earth. I find I mostly use salt to cover over and soak up red wine stains on carpets and for when I need a good scourer in the kitchen, but I rarely eat it, and food tastes much better without it. Salt has been linked to heart disease and the World Health Organization recommends only 0.2 ounces (5 grams) per day. In Britain and the United States we consume at least double that figure.

Avoid gelatine

Gelatine is one of those insidious ingredients that crop up everywhere. It is found in jellies, sweets, cookies, ready-made desserts, and cakes—especially those with glazes. The leftovers of animal gristle and bone make up the main ingredients, and you should always check the label for gelatine if you want to avoid animal products. Healthfood shops sell alternatives.

Eat healthier jam

Jams usually contain more sugar than fruit. Sugar was originally used to preserve the fruit so that jam could be eaten all winter. Mrs. Beeton wrote in 1859 that "sugar overpowers and destroys the sub-acid taste so desirable in many fruits." You can buy jams and preserves made from organic fruits without sugar, preservatives, colorings, and chemicals and refrigerate them. However, they are so tasty they don't even last long in the fridge.

Eat more honey

Bees work hard to produce honey, a natural sweetener that is better for us than refined sugars. If you can't give up sugar in your hot drinks, try half a teaspoon of honey as an alternative.

Use organic olive oil

Olive oil, and especially organic olive oil, is probably the purest form of cooking oil there is. It is recommended as the best and safest to use for salads, frying, and roasting. Sunflower and pure nut oils come next as healthy options.

Farm it

Shopping at a local farm market can give you more than just fresher food. In one market you can find hundreds of varieties of fruits and vegetables, not just the few dozen that your supermarket offers. This encourages real diversity, supports small farmers and especially organic farms, cuts down on the pollution of transportation and the middleman and, of course, the farmers get a fairer price.

Fight obesity

Obese people generally do not eat twelve organic apples to satisfy their hunger. They consume chocolate, sweets, potato chips, cakes, and all manner of sugars and carbohydrates. Those suffering from obesity may well be suffering from an illness making them eat more and more of this type of food. One theory is that obese people store up more chemical and pesticide residues in their fat and when they try to slim with conventional slimming aids, usually increasing their intake of chemicals, they get a toxic body-load too high for the liver, which can be damaged. Their body fights to reduce the toxic load to the liver by making the person crave more food. Chemicals and fat are an important issue for the 1990s and more research must be done. Food addiction self-help therapy is often a good way to come to terms with this problem.

Make your own potato chips

The processing of fresh potato into chips is a classic example of how food can be manipulated to cost more than its worth. A packet of chips needs energy, oils, fats, salts, and sugars, and it ends up costing 500 times more than the original potato. Why not make your own chips at home?

Learn to cook

It is not a joke. Many people just don't know how to cook and end up buying prepared meals with all the packaging, additives, and problems that processed food brings with it. As we spend less time cooking today, many cookbooks have been written with fantastic quick recipes, so you do not have to spend hours in the kitchen. Read a cookbook and try out the recipes.

Buy health-foods

Health-foods are unrefined, simple foods. They are usually pulses, grains, nuts, fruits, and vegetables that have not been processed and do not have additives or chemicals added. Less processing and refinement means that less energy is used in production and you get a healthier food. Health-food stores sell many varieties of what used to be staple foods. Some supermarkets have reintroduced health-foods and they work out being much cheaper than processed foods.

Eat more raw food

Raw foods, especially fruits and vegetables, are better for your body and also for the environment as they use little or no energy in their preparation. Raw food loses none of the essential vitamins and minerals that are lost in overcooked vegetables, for instance. You should always buy raw food in preference to frozen, canned, or processed varieties. Buy organic, but if you cannot, wash everything thoroughly before eating it.

Sprout food for health

Sprouted foods are simple to grow and extremely nutritious. You only need a few jam jars, clean water, and fresh seeds or pulses. The seeds give a volume of about eight times their weight and only take about three to five days to grow. Many raw energy recipe books give details of how to grow what works out to be an extremely cheap, nutritious food.

Make your own yogurt

Yogurt is important for our bodies because of the lactic acid bacteria it contains. It is simple and easy to make and you do not need the stabilizers and artificial sweeteners that are found in some store-bought varieties. It is often naturally sweeter than plain store-bought yogurt and you do not need to go to a lot of trouble to make it.

Eat more nuts

Many nuts have more advantages than you might think. They are a simple food, low in the food chain, and need no processing. This saves energy, pollution, and waste. Pumpkin seeds could help people with bladder-related illnesses; sunflower seeds have vitamins A, B, D and E and are good for headaches; all nuts give protein and energy that outshine meat-protein foods, help cut down on heart disease, and provide tasty snack alternatives to sweets.

Don't waste it

Millions of tons of vegetables and fruit are destroyed each year because prices had to be maintained at certain levels. Butter mountains, wine lakes, and tons of meats, vegetables, and fruits rot while people starve because our political and financial system cannot cope. This waste is not only unnecessary but cruel and environmentally bad. Campaign for a better system and don't waste food.

Pay more for your food

If we paid just 1 percent more for our exotic fruits, it could be possible to pass on 10 percent to the producer—the poor in developing countries. Pay the right price for your food to the right people.

Counteract "heavy" metals

By eating lots of certain types of foods with calcium, vitamin C, and traces of zinc, you can help absorb and destroy "heavy" metals that come into your body. Eating apples, carrots, and similar fresh, raw foods also helps as they are high in pectin. Toxic metals like lead, copper, aluminum, silver, and mercury are absorbed and taken into the body by a number of routes but mostly through contaminated food.

Don't buy your fruit from the street

Fruits and vegetables displayed on stalls in heavy traffic will have high levels of lead in them. Unwrapped or unpeelable fruits and vegetables will have the highest risk. Car exhaust fumes, general air pollution, and other contaminants get into food if it is left on display all day in the open air.

Discard the outer layers of vegetables

Cabbages, sprouts, lettuces, and similar vegetables will have pesticides and chemicals sprayed over them while they grow, but they will also have been subjected to further contamination and pollutants from the air. Remove the outer layers of these foods and wash vegetables very carefully before you cook them.

Don't cook with aluminum

Aluminum foil comes from mined bauxite, some of which is from destroyed rain forest areas. It is extremely energy intensive to produce, using nearly six tons of oil for every ton of aluminum. Too much aluminum in the body has been linked to memory loss, senile dementia, and Alzheimer's disease. Reduce your use of aluminum by not using aluminum foil when cooking foods. Avoid foods already packaged in aluminum. Do not grill foods on foil or put foods like potatoes in the oven wrapped in foil. Use a skewer instead.

Cut down on processed meats

Cured meats, like sausages, bacon, and hams, can contain added nitrates. Sausages often include the slaughterhouse products that are not allowed in uncooked meat products: the brains, feet, rectum, stomach, udders, testicles, and other parts of the animal. They are simply labeled as "offal." Water is usually the most important ingredient in processed meats, and complicated "rules" allow manufacturers to add vast amounts without declaring it clearly on the wrapping. Cut down on processed meat products for your health until manufacturers label their products clearly.

Spread a little less

Believe it or not, using less butter, margarine, and other fats on your bread, breakfast toast, and food can help improve health, as well as help the environment. Too much fat in your diet encourages the use of intensively reared livestock, especially cows. This takes its toll on the land, the animals themselves and, of course, on the food that is grown to feed them. By cutting down on fats you can reduce the risk of heart disease and of circulation problems.

Eat brown eggs

White eggs are produced from White Leghorn hens. They make up over 90% of America's hens and they are produced by a dozen major producers. By eating brown eggs, you can at least encourage diversification in the gene pool and encourage people to buy something different.

Drinking

Do not buy beer cans with plastic ring pulls

Next time you think of buying a six-pack of beer with plastic ring pulls, think again. The ring pulls are said to take 400 years to biodegrade, and, left as litter, they create problems for wildlife on land and sea. They throttle birds that forage near dump sites, and animals like weasels or otters can choke to death on them. Around the coast sea mammals and dipping birds have had their lives shortened by this handy little device. If you stop buying them, the brewers will have to stop producing them.

Don't drink and drive

Never drink alcohol and drive. Numerous campaigns have been run by congress and by local and national action groups to stop this foolish and dangerous activity. Twenty-three thousand seven hundred people die every year on American roads from drunk driving, and 650,000 are injured. In Australia the introduction of random breath-testing lowered drink-related accidents by one-third. European countries have followed suit, and in the United States Mothers Against Drunk Driving forced many states to ban open drinks in cars and set up road blocks to catch drunk drivers.

Refuse hormones in drink products

During 1989 a row erupted over the sale of milk that was treated with the hormone bovine somatotrophin (BST), a hormone that can increase a cow's milk yield by up to 20 percent.

The long-term effects of hormone treatment on cows are still unknown, but in nearly every case tested so far the evidence has come down against their use. What is worse is that while testing has been taking place the milk has passed, unlabeled, into the general supply. Contact your milk supplier and complain. Refuse milk products containing BST.

Opt for chlorine-free cartons

Most fruit juices and milk are sold in cartons made of paper mixed with bleached paper, foil, and polythene. Chlorine bleaches are used to make the paper white, causing serious environmental problems at paper mills when the waste is flushed away. Even more worrying are studies conducted by the Canadian and New Zealand governments showing that dioxins, the chemical by-products of the paper bleaching process, are found in the paper and can migrate into the milk or juice itself. Some companies are now producing chlorine-free cartons, which are brown, but you will be better off using bottles and recycling them.

Try real cider

Cider is a traditional drink, brewed originally from thousands of varieties of apples. Today many ciders are made from apple concentrate imported from countries that place few restrictions on chemical fertilization. The Alar pesticide scandal of 1989 pointed to real worries about the use of such chemicals on apples, as they may well be cancer causing. Alar has been banned in the United States, but the British government has given it the okay. Traditional cider is made without chemicals, so look for them and support them whenever you can.

Use loose tea

Tea is popular all over the world, but nowadays we have created another use for paper in that convenient device, the tea-bag. Millions of tons of chemically bleached pulp are mixed with wet-strength resins, including urea formaldehyde, to make tea bags. Tests have found dioxins, including the most dangerous, TCDD, in measurable quantities. Whenever possible, save paper, save on chemicals, and buy loose tea.

Don't drink yourself to death

Alcohol can be a killer. It destroys the liver and kidneys and the central nervous system, and it affects almost every faculty imaginable, from intellect to sight. Thousands of people are killed and injured on the roads each year as a consequence of alcohol-related driving accidents; thousands more die from related illnesses; and families have been emotionally destroyed by alcoholism. Alcohol can be addictive and should be treated as a drug at all times. It is to be enjoyed in moderation and should be produced without chemical interference or animal testing.

Try an herbal tea

Herbal teas are usually made from locally grown plants and are a delicious alternative to caffeine-laden coffee. There are now wonderful teas on the market, such as strawberry, chamomile, and peppermint, some of which have medicinal properties, too. Avoid those that are heavily overpackaged, and if your favorite is, write to the manufacturer about it. The best option is, of course, to make tea from your own herbs.

Tea is a treat

North Americans drink on average 154 cups of tea per year per person. It is generally thought to be a good drink for you and the environment. There are no chemical additives, preservatives, or sweeteners, unless you add them yourself in the form of milk and sugar. Tea plantations have a bad reputation when it comes to paying reasonable salaries to their workers, though. We think of tea as a cheap drink. If it were not so cheap, we would think twice about throwing it down the sink so readily.

Why not buy your tea from cooperatives that ensure that some of the profits go to the workers?

Reduce your caffeine intake

The amount of caffeine in coffee is much higher than in tea, but coffee is fast becoming the number one hot drink in affluent countries. Environmental problems are caused chiefly by the pesticides used on the coffee plantations. Caffeine has been linked with high blood pressure and heart diseases; doctors recommend cutting down on your caffeine intake. If you can find organic coffee, buy it, but try to reduce the amount of coffee you drink.

Reuse your coffee filters

Although some of the big coffee-filter manufacturers have been quick to produce chlorine-free coffee filters after the Women's Environmental Network and Greenpeace protested about the environmental problems caused by bleaching them white, we do not need to use expensive paper filters at all. Rinse out a metallic or cotton reusable filter.

Reduce your milk and milk-product intake

Milk and milk-based products are popular food items. But milk is one of the most sensitive and allergenic foodstuffs. The majority of the world's population does not drink milk or eat milk-based products at all, and those people manage to get all the nutrients and calcium they need to survive. The intensive rearing of cattle for milk production uses valuable land, and the cattle eat much of what could give us an alternative source of protein and calcium. Milk has been implicated in over 70 percent of cases of allergic skin reaction and nearly 90 percent of asthma cases. If you have breathing or sinus difficulties, try giving up milk completely and you will notice the difference. Use milk as a special additional food and not as a staple food,

and you will be in better health. You will also reduce the need
for intensive cattle rearing.

Oppose animal abuse when testing alcohol

Animals are literally forced to be inebriated to test the effect
that alcohol has on humans. During 1987, 3,746 animals were
"tested" with alcohol. There can be no justification for this
cruelty.

Recycle your PET

PET bottles have become a household wonder. They are lighter
than ordinary glass bottles and if accidentally dropped, they
will not smash, which saves on waste. PET stands for poly-
ethylene terephthalate, and no data exist to suggest its toxicity.
We do know that these bottles are not biodegradable, though
they can be recycled. As there are at the moment few facilities
for recycling the billions of PET bottles produced, we should
send them back to the manufacturers and ask them to recycle
the bottles until they can establish specific collection points.

Recycle your can

We buy a staggering 70 billion cans every year in the United
States for soft drinks. Aluminum cans can now be recycled in
many areas, and in the next few years we should see the small
number of cans we recycle rise to a more environmentally
acceptable level. When you buy your next canned drink, remember that over 80 percent of the cost is for the can itself!
Avoid buying things in cans, recycle the cans, and opt for draft
or bottled drinks (as long as you aim to recycle your bottles
too). Recycling cans saves up to 95 percent of the energy cost
of producing them.

Drink safer water

The water that comes from our taps these days may be far from
safe. It may contain nitrates, lead, aluminum, pesticide, fer-

tilizer residues, and a host of other chemicals, some of which are natural. Chlorine, used in water-treatment plants to kill bacteria, forms trihalomethanes if mixed with organic material such as peat from moorland or lowland rivers. These are thought to cause cancer of the bowel and bladder, and new studies have linked increases in leukemia with chlorinated drinking water. Until we can all have safer drinking water it may be best to filter your water, using approved filters. The only goal we should have, however, is safer, cleaner water to drink from our taps.

Bottled chemicals

A bottle of mineral water may not be a safer product. Tests of many bottled waters have identified bacteria, pesticides and nitrates. Add to this the environmental cost of producing non-returnable, nonbiodegradable, plastic and glass bottles. Fizzy mineral water is a good alternative to alcohol and if you buy a clear glass bottle it can at least be recycled, but beware—there are not approved standards for the production of such water yet.

Choose organic and animal-free wines

Organic wines are on the increase as discerning drinkers are rejecting wines treated with sorbic acid, sulphur dioxide, ammonium sulphate, diammonium phosphate, and up to fourteen applications of pesticides. Animal and mineral extracts are used in the production of wine, including edible gelatine from bones, isinglass from the bladders of fish, casein and potassium caseinate from milk, albumin from egg, dried blood powder, tannin from wood, kaolin, and clays. Many organic wine growers claim that you will not get a hangover if you drink organic.

Brave a better beer

Beer used to be made from four simple ingredients: hops, malt, yeast, and water. These days we have no way of knowing what exactly is in our beer, and the biggest chemical ingredient, ammonia caramel, is prepared by the controlled heat treatment of carbohydrates with ammonia. This is not considered to be toxic, but vegetarians or vegans will want to avoid beers that have been force-fed to animals for testing and those that use isinglass finings, which come from the bladders of fish. Buy beers that have been filtered without using animals.

Do not buy paper or plastic cups

The waxed paper cup was invented in 1907 by an American, Hugh Moore. He produced it to sell pure chilled drinking water, and it soon replaced the communal drinking ''sipper'' as a symbol of good health. The idea took off, and now everything from water to ice cream is sold in it. Every year we use billions of disposable paper cups around the world as our parties and celebrations have moved away from glasses to throw away paper or plastics. The paper cup comes from paper bleached with chlorine, which inevitably causes dioxin contamination around the paper mills, leading in turn to pollution of the water system. Plastic cups are made from petrochemicals, and they too have their own environmental impact. Use strong glass wherever you can.

Go for 100 percent juice

Fruit drinks are not always what they seem. Most juice drinks that we buy contain as little as 15 percent fruit, and they usually cost as much as the real thing. Some contain colorings, flavorings and large amounts of disguised sugars and artificial sweeteners. Go for 100 percent juice whenever you can. Try squeezing an orange (in its own biodegradable wrapper) and you'll be sure of getting the real thing. Campaign for better labeling.

Complain about your drinking water

The tap water that millions of people are drinking may not meet legally binding standards. Write a letter to the people in charge of tap-water quality in your area. By law you are entitled to ask for details of contaminants exceeding the legal limits in samples and to make a formal complaint to the water authority and the government.

Presents and Parties

Don't fly away with balloons

Balloons are used by campaigners and party goers alike, yet most of us are not aware of the environmental problems they may cause. If they end up in the ocean, for example, they may kill the fauna that swallow them. Sea turtles are at particular risk, as once a balloon has lost its color, it looks just like jellyfish, their favorite food; but the larger sea creatures are also threatened. In 1985 a 17-foot-long female sperm whale died of starvation off the New Jersey coast because the balloon she had ingested blocked the valve connecting her stomach to her intestine. Foil balloons are not biodegradable. Rubber balloons are, but they have clips and strings that also cause problems. They all must be disposed of carefully.

Buy a potted Christmas tree

Your Christmas tree dies the minute it is chopped from its roots, but if you buy a locally grown tree in a tub, with its root, you may replant it in your garden after the festive season. Millions of trees could be spared in this way. However, the real green alternative is to decorate the large houseplants in your home.

Buy green books as presents

There are now many green books on the market. With hundreds of titles to choose from, you will be able to select valuable and informative gifts for everyone.

158

Golden problems

One of our most precious metals, gold, is bought and sold all over the world. It is central to the functioning of our financial institutions, and its price has changed little over the years. Much of the gold on the world market comes from South Africa and the USSR, but gold can also be found in the forests of the Amazon. Gold processing uses and produces cyanide and mercury, which are responsible for water pollution and are particularly toxic to fish and aquatic life. In South Africa bad working conditions, pollution, and poverty shorten the lives of many of the black gold miners and the Anti-Apartheid Movement has called for the boycott of South African gold. In Brazil, where there are 50,000 gold diggers, gold mining is destroying the fabric of the Amazonian rain forests and putting the lives of the Yanomami people at risk. Think carefully before you buy your next gold present.

Wrap up presents with something different

Instead of wrapping your presents in paper that has been bleached, why not think of something a bit more unusual? Recycled paper in natural colors, banana paper, hessian, muslin, rush or bracken, wool or cotton are alternatives that make presents look really special.

Do not buy ivory

There are only 600,000 elephants left in Africa. In 1979 there were 1.5 million. Elephants are killed by poachers using either dynamite or bullets to provide us with exotic ivory trinkets, carvings, and piano keys. Their tusks are bought whole by Japan, Hong Kong, and Taiwan and, after being cut, are sold around the world. Sustained campaigning by groups like the the Environmental Investigations Agency has brought about a moratorium on the ivory trade that should be strictly enforced. Do not buy ivory presents, wherever you are. Boycotting the market in such trinkets will help to protect one of our most majestic mammals.

Do not buy kangaroo gifts

The Australian government allows the killing of 2 million kangaroos every year. It says that their number needs to be kept down and that useful products can be made from their skin and meat. Millions of dollars are made around the world out of the sale of kangaroo-hair toys, kangaroo-skin car-seat covers, gloves and rugs, and kangaroo meat is used in pet food. If we stop buying such items, we may find that the kangaroos do not need to be shot at all.

Be careful with fireworks

Fireworks can be fun for humans, but they are terrifying for animals. The ASPCA advises cat and dog owners to give sedatives to nervous pets when fireworks are to be let off near them. The chemicals in the fireworks are not only a danger to people but their production constitutes an environmental hazard. You could have just as much fun baking potatoes and chestnuts on an open fire with friends.

Serve organic food at parties

If you cannot afford fresh organic fruit and vegetables all year round, or find them difficult to get hold of, serve them as a special treat at Christmas and birthday parties and on other special occasions.

Save the croc

Every year 340,000 crocodiles are killed for their skins out of which expensive shoes, handbags and briefcases are made. Only a few farms exist, so most crocodiles are captured in the wild in Papua New Guinea, Venezuela and India, and in the Nile and the Pacific. The crocodile-skin trade is worth $68 million: one average skin is worth $200. The biggest markets are Japan and France, but other countries import crocodile skins, too. Do not buy items made out of the skin of wild animals like the crocodile.

Boycott battery-operated presents

Buying presents at Christmas and for birthdays is difficult; there is always a huge selection of new and exciting games and products available. One way of helping to protect the environment is to stop buying battery-operated toys. The energy spent making a battery is fifty times greater than the energy you will ever get from it—a waste of resources on nonessential goods. The batteries themselves contain toxic and dangerous substances, including cadmium and mercury, which may leak if the batteries become damp. Even "greener" batteries may contain zinc, carbon, mercuric chloride, and alkaline manganese.

Don't forget the walrus

Now that concern over the mass culling of elephants is increasing, the next mammal on the hit list for tusks is the walrus. Greenland and Canada have access to walrus tusks: the only way to make their sale unacceptable is to boycott it and to campaign to stop the trade now.

Support local flower growers

Flowers make lovely presents to receive and give: greener flowers are even nicer! Those imported from other countries are more likely to have been sprayed with pesticides than those bought from small-scale local growers, and they have also been flown from overseas at great expense and energy. Also bear in mind that recent concern has been voiced over the health conditions of women in Honduras who earn their living growing carnations to send to Britain. So if you have a garden, grow your own flowers; if you don't, start asking about chemical-free alternatives.

Always serve nonalcoholic drinks

Do not encourage drunk driving; at your parties always have a choice of interesting and tasty alcohol-free drinks. You can make a hot apple punch without alcohol and serve exotic fruit bowls, shakes, and juices. Nonalcoholic wine has a nice flavor and, as nonalcoholic beer is the biggest growth area for the brewing market, there are now plenty of brands to choose from.

Recycle your Christmas tree

On January 6 every year 35 million Christmas trees are thrown away as rubbish in the United States. But these days new schemes are emerging whereby, for the cost of $1, you can recycle your tree: it is chipped and made into a mulch that is then either sold or spread in public parks. Some communities in Texas have been using the trees as a base for dunes to fight beach erosion, which saves them from being dumped in overflowing dumps.

Organize a green raffle

Raffles, sales, and prices in national and local events play an important role in the community. You could start to influence any committee, organization, church or group that you belong to next time you decide on a raffle as a fund-raising idea. Why not have green prizes? Reject the car, chemical products, the rain forest-wood furniture. Go for organic hampers, bikes, camping trips, cruelty-free cosmetics, and recycled paper kits.

Greening your hampers

Picnic hampers have always been exciting presents to receive. These days you can put together a more than acceptable green hamper that includes organic wines and champagnes, organic fruit and vegetables, and even sugar-free alternatives to sweets and desserts. Add some real holly and a picnic basket from your favorite charity shop, and you have a very special gift.

Free-range it at Christmas

Christmas and Thanksgiving are traditional occasions for serving turkey. In the United States, 240 million turkeys are munched each year, 99 percent are factory farmed. Free-range and organically reared turkeys are the choice for the green meat eater. They cost more, but the meat is apparently more succulent and tastier.

Investment

Choose your home carefully

The purchase of a new home, of whatever size, will probably
be the biggest expenditure the average person ever makes. You
should consider it as a long-term investment. If the house you
are planning to buy is being built, you may be in a position to
insist that the builders include energy efficiency and even the
use of nontoxic materials in the specification. Many of the
larger building companies are beginning to sell greenhouses.
Don't be conned into paying too high a price, however. Some
companies have been building energy-efficient houses for years
at no extra cost to the buyer, so shop around and choose your
home carefully.

Ethical friendly unit trusts

If you have money to invest and you want to buy unit trusts,
there are some that operate according to strict principles that
are based not just on environmental considerations but on a
range of tenets that the average "green" would endorse. You
can avoid businesses associated with the arms trade, alcohol,
gambling, and animal exploitation, oppressive and fascist regimes like South Africa, or those that have bad reputations as
employers, and still get value for your money.

Support businesses that give to charity

Charities are vital links for the community. They spot trouble
much faster than governments; they generally support the ordinary person rather than corporations or politicians; and they

usually offer extremely good value for money. More and more corporations are giving charitable donations to environment groups, social services, the arts, and education. If you support such companies you will be reinvesting in charity. Beware of paying twice for a service, however: the government, not charitable bodies, should pay for essentials.

Give 5 percent of your income to a deserving cause

In the United States the average amount given to charity is 2.4 percent of an annual income. Most of us could live on 5 percent less. Try it for six months (you can get tax deductions for charity donations). Get to know the groups and organizations that you want to support.

Don't invest in arms

The great nations of the world spend $1 trillion on defense a year. The folly and waste that goes into servicing a system continually prepared to kill, while millions starve, is inhumane and barbaric. There is only one planet. Nuclear arms, nerve gases, bombs, and rockets kill and maim, and they have the power to destroy our ecosystem. We should be putting our spare cash and scientific energy into solving our environmental crises instead of creating new ones. Don't invest in war: boycott companies that manufacture armaments.

Write to the World Bank/IMF

Every second over 1 acre of tropical forest goes up in smoke. Since 1960 we have lost over 40 percent of the world's tropical forests, the habitat of the world's most beautiful and rarest plant and animal species and the home of approximately 150 million indigenous people whose lives depend on them for survival. Through taxes and government contributions your money goes to the World Bank/IMF: ask the World Bank/IMF in Washington, D.C., to swap the debt that is owed to them by these countries for the protection of the rain forests.

Stocks and shares for the environment

A good broker will be able to advise you how to buy stocks and
shares in green companies. Your investment will support them
and, with luck, can earn you interest at the same time. Companies such as the Body Shop have seen their share price rise
by over 200 percent in just a few years as more and more
bodies have been willing to invest in green businesses.

Organize a credit union

Cooperatively run credit unions can make a real difference to
the poor. Loans may attract as little as 12 percent interest,
compared with the normal 18–19 percent for credit cards and
up to 1,000 percent for a highly profiteering loan shark. Credit
unions operate by sharing money according to need, on the
principle that profit isn't necessarily their most important goal.
The interest pays for the union to operate rather than for profit
for companies or individuals.

Tell your bank what you want

Why do banks and building societies continue to send us mailing after mailing about new loans and new insurance plans and
invite us to borrow more and more? What a waste of paper!
Ask your bank to take your name off their mailing list; ask it to
provide checkbooks and statements on recycled paper, as there
isn't any justification for bleached new paper for these. And
what about facilities for disabled people? Banks hardly go out
of their way to accommodate even the able-bodied—they must
be a nightmare for others. Don't forget: the money you put into
them is what they live on. You are the customer, and you are
entitled to ask for what you want.

Change policy from the top

Ask financial businesses to take the environment seriously.
Banks, building societies and loan companies could all be
asked to account for their environmental performance. Ask

them to present an environmental report in their annual report to shareholders. Ask them to name one person on their board of directors who would be responsible for the environment.

Ask for environmental audits for loans

Ask your bank or building society to consider environmental audits for new loans, then a list of acceptable criteria could be drawn up and preference given to loans for less damaging or polluting schemes and businesses. Lending institutions are responsible for giving away billions every year, most of it for new homes, but a large percentage for home improvements, cars, and new businesses. Just think what a difference it would make if you had to answer a question on your loan form about the effect your loan would have on the environment.

Ensure a greener future for your children

Give your children a greener future by investing your savings for them in an environmentally conscious fund. Ask your bank or building society to suggest one: it could be an investment for more than your child alone.

Write to your bank about debt in the Third World

Write to your bank about Third World debt. Is the manager of your bank doing anything about Third World debt, and is he prepared to write off the debt in favor of environmentally sustaining projects? Most banks are involved in developing countries, and they are sensitive about what you think of their reputation.

Safeguard your possessions

There are now green insurance funds. They can provide you with insurance coverage for your house and car that uses an ethical investment service. They work along the same lines as other ethical investment programs: your money will not be invested in arms manufacture, tobacco, military dictatorships, South Africa, or the support of environmentally destructive programs. Ask your insurance broker to find out for you.

Bequeath your money to the future

When you write a will or bequest, think about leaving some of your money to environmental groups and organizations. This could make a huge difference especially to smaller organizations, which don't attract the attention of larger groups. Many groups will advise you about how to write a will if you don't know where to start.

Guarantee an insurance that is ethical

Life-insurance funds are common for partners when they buy property jointly: if one partner should die, the life insurance will cover the other partner financially. Ask your life-insurance broker to find you a green fund. This will mean that from the start your money will be invested in sound businesses and environment-conscious trust funds.

Retire in greener pastures

Many jobs include pension funds as a matter of course, and billions are invested in such funds by powerful brokers around the world. You may argue that your money should be invested in ethical funds for you. The law of fiduciary trust, however, means that your company is required by law to provide its investment members with "the best available rate of return," which sometimes cuts out some of the ethical funds, especially the newer ones. But where you can influence your company, do so. You can transfer your funds, and pension advice schemes will readily give advice on how to do this.

Choose a responsible way to share your wealth

If you inherit or own some money and you want to share it with others, choose an issue that worries you most. It's worth spending time talking to people and informing yourself thoroughly about the matter before you make your final decision. There are groups of people who meet and share ideas about giving money away responsibly in the United States and Britain. You could help to influence the future by supporting groups and individuals generously.

Order a greening loan

Pressurize your bank or building society into lending you money to use for greener projects—fitting your home with all the latest energy-saving equipment, for example. This investment would be repaid within a few years and would add to the value of the home, so it would be well worth it. If lending agencies were greener, we might see this sort of loan being given in preference to loans for other house improvements, and fewer new power stations would be needed.

Green up your bank

Is your bank investing your money in environmentally sustainable projects? Some banks have been exposed for investing, directly and indirectly, in programs that destroy rain forests in Brazil; others have invested in dam projects that have destroyed local ecosystems in Asia and South America. Before you open a new account with a bank, ask it to give you an account of its practices. Choose to open an ethical and environmentally sound deposit account that invests only in cooperatives.

Take a green credit card

If you have to use a credit card, why not use it to support one of your favorite groups? Credit cards are becoming increasingly popular with larger organizations as a method of collecting funds. One of the main problems they encounter, however, is a moral one: should they support spending and consumerism? If you are able to live with that dilemma and you need a card, then do at least get a green one.

Get advice

A growing number of ethical investment services offer advice on specific investments. They should be approved by appropriate professional bodies, so do ask before you take advice from them. They will be able to tell you about the ethical or socially responsible funds available on the market, and they will advise about green pension funds, personal equity plans, and investment funds.

Pay bills by automatic withdrawal

Ask your bank to pay your bills regularly by automatic withdrawal. Think of all the paper you will save.

GREEN WORK—
GREEN LEISURE

Green Workplaces

Develop an environmental policy

Over 80 percent of businesses are small ones, employing between twenty and thirty people. It is easy to develop an environmental policy in small workplaces or offices. Start raising general environmental awareness by asking questions, and do not accept "no" for an answer. Saving the environment can very often save your company money, by using energy efficiency programs or recycling schemes.

Begin with an environmental-impact assessment

Environmental-Impact Assessments (EIAs) ensure that, when large projects are at the early stages of planning and feasibility studies, possible effects on the environment are taken into account. All businesses should aim to limit environmental degradation from the onset. At present, certain businesses in Europe including crude-oil refineries, power stations, installations for storing and disposing toxic and dangerous wastes, trading ports and inland waterways, chemical installations and steelworks, are required by an EEC directive to carry out EIAs. All businesses should carry out an Environmental-Impact Assessment at the planning stage. It allows better choices to be made, and it is still early enough to modify the original scheme.

Have a smoke-free office

Millions of people smoke in workplaces across the country. Not only do they inflict a series of health risks on their colleagues but they are also directly responsible for causing pollution. Tobacco production and manufacturing processes include heavy use of pesticides on the tobacco-plant leaves, the bleaching of cigarette papers, the pollution of rivers and the air. Smokers also pollute their own bodies, running all sorts of health risks that cost your company money in sick pay: a report by the Dow Chemical Company showed that smokers had 80 percent more sick leave than nonsmokers. Smoking is also a fire hazard. Negotiate a smoke-free workplace and help smokers kick the habit that may otherwise eventually kill them.

Ventilate your photocopier

Ozone can be an indoor pollutant, especially if you work in a room without adequate ventilation near photocopiers that are switched on all day. Ozone can be damaging to both the mucous membranes and the eyes, and it is also thought to cause persistent headaches. Use your photocopier with care, and always ventilate rooms.

Is your building sick?

Artificial lighting, open-plan offices, and artificial ventilating systems can create an alarming new illness: Sick Building Syndrome. Sick buildings have been noted in modern office buildings that have artificially created atmospheres. Space-age office machinery has added to the problem, together with chemical fumes, electrostatic interference, and electrical risks. It all adds up to a pretty unpleasant atmosphere for workers, who complain of general malaise, lethargy, headaches, and the like. This affects productivity and it is therefore in your employer's interest to improve conditions. Check your building. If you think it is sick, take measures to make it better. Contact health and hazard organizations for help.

Don't type so fast!

New technology might save paper, but it can also make you ill as your body attempts to keep up with the speed of the technology itself. Work-related upper-limb disorders or repetitive strain injuries (RSI) affect workers who use electronic keyboards and supermarket bar-code readers. Injuries to the tendons, muscles and joints can result, and this can eventually lead to the loss of use of hands or arms. Operators who are paid piecework rates are most at risk. A typist will average 10,000 keystrokes per hour, but some workplaces offer incentives of extra pay or holidays, and organize competitions, thus pushing typists into producing an excessive 27,000 keystrokes per hour. Computers that can monitor performance are putting extra pressure on workers to go faster and faster. Make sure your chair is angled correctly in front of the VDT, do not agree to excessive speeds for typing and take a break as often as possible.

Get a shield or filter for your VDT

Computer terminals and VDTs have caused problems for workers, as it has been discovered that they emit low-level radiation. They have also been linked to increasing numbers of miscarriages, and other illnesses have been reported, including headaches and skin rashes. I fitted a filter screen on the office computer and the rashes and ''hot'' faces were gone within a day. Always take a break after an hour spent working on computers, work for no more than four hours a day in front of a screen and contact trade union and health groups, who have information booklets about the risks.

Do you really need sprays to clean your telephone?

Shared work telephones could possibly spread bacteria or bugs, but cleaning chemicals may be more toxic than your own germs. Aerosol cleaners, even now with ''soft'' CFCs or hydrocarbons (another greenhouse gas) are mixed with highly toxic and flammable disinfectants which should not be inhaled in excessive amounts. It is far safer to wipe the telephone with a clean, damp cloth, and it is cheaper, too.

Cut the neon lighting

Neon lighting was discovered in 1898. Neon is an odorless gas that can be illuminated in a tube. Added powders alter the state of the gas, producing the full spectrum of colored lights we now know. Some shops use different types of high-tech laser and computer lighting to sell their wares, but all of them are wasteful of energy and produce tons of pollution in the form of carbon dioxide and other greenhouse gases. See if you can do with less neon.

Lighting for health

Our eyes suffer from continuous strain in office situations. Reading from bad photocopies, illegible writing, and carbonated papers all put strain on the eyes, but this causes far less strain than the lighting itself. Artificial lighting in offices attempts, without much success, to replace natural lights. White fluorescent strip lights can have a constant flicker and can cause headaches, eyestrain, and concentration problems. Desk lamps offer some flexibility and create less glare. Using energy-efficient lighting means that you use twice the amount of light for less money. You can have both a good clear light for work and save energy, too.

Have an energy-efficient workplace

Saving energy in your workplace will save your company money but can also contribute to reducing global warming by reduced carbon dioxide emissions. Don't leave office lights on at night. Use compact fluorescent bulbs, which save money and have low wattage. Turn off machines not in use, especially computers and photocopiers. Install efficient heating equipment. Make someone responsible for energy saving in your office.

Ban the vending machine

The hot-drinks vending machine was brought into offices in the last fifty years, mostly to replace the expensive tea lady in larger firms and companies. It serves synthetic drinks, powdered milks, coffee and tea, mostly in plastic cups. The few minutes time a vending machine saves staff are lost on energy costs alone. The average vending machine guzzles electricity—it may be the largest single user of energy in the workplace. Using mugs and low-energy water urns might be a better option and give you tastier drinks.

Static-free floors

Increasing numbers of complaints have been recorded in workplaces about a strange new "bug" found in artificial carpets. Synthetic backings (foam underlays may contain CFCs) and plastic coatings encourage carpets to attract static. This in turn has been "jumping" up and "attacking" workers on the legs. The buildup of static electricity is further encouraged by the use of VDTs and other machinery. A natural carpet with hessian backing will reduce the static immediately. Using little or no synthetic chemicals in the coatings and as cleaners also helps.

Use long-life equipment

It is obvious that office equipment that is used by many people needs to be hard-wearing to be economic, but the better the equipment you buy, the better for the environment. Use materials like metal filing cabinets instead of paper or cardboard ones. Buy from secondhand shops, which is also a form of recycling—and cheaper, too!

Catch the cheap desk out

Cheap veneer desks can harbour numerous environmental hazards. They have chipboard interiors which, although chipboard is rarely made from rain forest wood, it is usually bonded with formaldehyde glues that can cause problems if they seep out. The veneers are likely to be from rain forest–wood sources, however, unless they are plastic laminates, which attract dust and dirt. Solid wood desks from sustainable sources are the best to use. You can even have them waxed and varnished with organic materials, for easier care.

Don't demand a rain forest–wood desk

Most of the tropical hardwoods come from the rain forests. This means that your new desk, and the chair you sit in, may well be from a rain forest source. Check out where the wood used in your office furniture comes from. Avoid hardwoods, especially if you can't discover the country of origin. Use the Friends of the Earth *Good Wood* Guide to help you.

CFC-free office furniture

It is not just aerosols that contain CFCs—chemicals that destroy the ozone layer. The foam used for most office chairs contains these chemicals. Check on this with your supplier. The chairs should also conform to strict fire-safety measures, as the burning foam can give off toxic fumes.

Audit your suppliers

You get what you ask for! Why not start auditing your suppliers? Do they stock recycled paper? Water-based pens? Energy-efficient lighting? The equipment and furniture that you buy to keep your workplace going need not be environmentally damaging products.

Don't use envelopes with plastic windows

Envelopes with plastic windows are the most environmentally unfriendly envelopes available. They are not recyclable and are likely to be made of virgin paper. Ordinary brown envelopes are mostly made from recycled pulp. If you have started to recycle paper in your office, you will find that window envelopes are not acceptable. Encourage your stationery buyer and your mail department not to use them—they are more expensive anyway.

Disposable writing

Every day trash cans in the United States contain 4,383,562 disposable pens. Around the world, the throwaway ballpoint pen has become a familiar part of our lives. Made of plastic, nonbiodegradable and rarely used right until the ink is finished, they are an environmental hazard that also wastes precious fossil fuels. Try using a refillable pen, cut down on disposables, or, if you must use them, don't throw them away until they are completely finished.

Correct your correcting fluid

Have you ever wondered what is in that lovely white correcting fluid you use for your typing errors? The main chemical is 11-trichloroethane, also known as methyl chloroform, a toxic and irritant chemical that depletes the ozone layer and stays around for a long time in the environment. It is particularly dangerous when burnt, as it forms phosgene gas. Look for water-based whiteners that say specifically that they have removed trichloroethane. If you can't find a good alternative, cut down your use of them.

Save paper

It is estimated that businesses in the United States use over 1.6 trillion sheets of paper each year; around 13 percent of these are unnecessary photocopies. A waste of several million trees. Add the energy bill to that and you will see that the photocopy machine seems to have created a wasteful problem. Cut down on the amount of photocopies you make, reuse paper written on only one side for notes, and always keep a bin next to the copier in which to place spoiled sheets for recycling.

Use clear glues

Adhesives, sticky tapes, and other glues are big business. Glues are a very necessary part of the office, but some are more toxic than others. Glue production itself can lead to environmental problems, as the volatile chemicals can escape into waterways and river systems. Glues can contain phenol, vinyl chloride, formaldehyde and epoxy resins, naphthalene, and other harmful chemicals. They are often addictive and volatile. Avoid them as much as possible. Stick to white glues and simple-formula woodworking glues whenever possible.

Don't use colored paper clips

Paper clips are a small but necessary item in the office. They can be recycled and reused many times, which makes them valuable and cost-effective, but say no to the colored paper clips that are dipped in bright plastics. They can contain cadmium, a toxic heavy metal.

Use water-based and felt-tip pens

Graphic artists are among those who will use colored pens most these days, but all offices and workplaces are using increasingly sophisticated means of communication: wipe-clear boards, notices, documents that need sections highlighted need good quality coloring pens. Permanent-ink markers contain solvents and chemicals like phenol-toluene and xylene. They are moderately toxic if swallowed and must be used with care in well-ventilated rooms. Prolonged exposure to these chemicals can cause dizziness, and toluene may have effects on the central nervous system that could be dangerous. Opt for water-based markers whenever possible.

Run down the battery use

When manufacturing batteries, up to fifty times more energy than the batteries actually produce is used. Offices and workplaces around the world regularly use batteries to operate calculators, clocks, and machines. Use mains-operated equipment, solar-powered calculators and hand-operated machinery whenever possible. If you must use batteries, try "green" rechargeable ones, which last up to 500 times longer.

Check your continuous stationery

Computers were hailed as the new technology that would rid us of unwanted paper forever. Waste was going to be a thing of the past. Unfortunately, it never really worked like that. Computer stationery ended up making us waste even more than we used to. Machines using sprockets for continuous stationery seem to be the worst, as there is often no way to stop them when in midflow. Collect your unwanted computer stationery in a bin near the computer; it is the most highly valued for recycling. Use single-sheet feeders whenever you can, and cut down on the numbers of sheets that are printed out whenever possible. Recycled computer stationery is available.

Use recycled soft-paper products

Toilet paper and paper towels should not be made from paper pulp from new trees, nor should the paper be bleached. Recycled paper products save your company money and protect the environment at the same time. See if you can cut down on the numbers of paper towels that are used, and switch to recycled paper products whenever possible.

Get rid of the plastic mailout

Magazines and newspapers in plastic bags are being mailed to subscribers all over the world. Literally millions of these bags are used each day. If your workplace sends mailouts with plastic wrappers ask your employer to return to recycled paper wrapping. These flimsy plastic bags are not biodegradable and are estimated to last for 400 years in landfill sites. Recycled paper, on the other hand, can be recycled again and represents a better use of resources. The publishers of American maga-

zines like the *New Yorker* and *Mother Jones* have already
bowed to public pressure and concern about the environment
and have reverted to paper once more. If you receive magazines in plastic mailouts, write to the publisher.

Avoid Styrofoam coffee cups

Styrofoam cups, together with Styrofoam plates and containers
from the canteen, should be avoided. They are made with
CFCs, which destroy the ozone layer, and they cannot be recycled. Bring in your own mug and ask the canteen to switch
to recyclable paper cups.

Reduce noise at work

Noise can be a pollution problem. It can damage the ears and
add to stress at work. Machines clicking, vibrating and hum-
ming, telephones ringing, typewriters and computers clattering
and bleeping all make a noisy environment for the average
office worker. Add that to a machine in the background and
you could have a real problem. Excessive noise affects your
senses; it contributes to headaches, loss of sleep, fatigue and
stress. Hearing damage begins at about 75 decibels of noise if
continuous, and yet most factories regularly operate at above
90 decibels. Even open-plan offices can reach 80 decibels. Use
engineering controls to reduce the noise of machinery, fit
acoustic controls like thick carpeting, heavy fabrics, and plants
which "absorb" noise. Consider soundproofing to reduce the
exposure of workers to noise pollution.

Beware dangerous chemicals

Industrial chemicals are being used in workplaces every day.
At least 2,400 are suspected of causing cancer, and many more
have unknown properties and effects. Office workers are exposed to powders, dust, gases, and liquids from machinery and
equipment. Factory workers are exposed to even more chemicals in large industrial settings and workers are often ill-

informed about their effects. Hazards are likely to increase as more machinery is used. It is your employer's legal responsibility to provide a safe working environment and healthy working conditions for you. Always use the minimum quantity of each chemical, ask about their specific properties, ensure that you have good ventilation and keep a careful eye on the health effects of any chemicals you may be exposed to. Cutting down the number of chemicals used in your workplace cuts down the production of these chemicals, and therefore the amount of damage they do to the environment, as well as to you.

Don't test fire extinguishers with halons

Halons are used in certain types of fire extinguishers, particularly those for use on computers and electrical equipment, because they can smother a fire with a gas and without destroying the computer. Although Halon 1211, the chemical usually used in the fire extinguisher, is nontoxic to humans, it is also a serious ozone depleter. Dr. Joe Farman from the British Antarctic Survey believes halon now causes some 14–30 percent of the ozone depletion over the Antarctic. And 95 percent of the emissions actually come from testing the extinguishers to identify leaks, from accidental releases, and from losses from old equipment. This is one place where we can cut emissions radically and immediately. Check that the extinguisher you use is halon-free. Air-pressure testing for leaks is an alternative to releasing these dangerous chemicals.

Ban aerosols

Aerosols are used in cleaning, graphic design, paint-spraying, and a host of other office and factory uses. In every case there is a friendlier alternative to this wasteful packaging. Hand pump sprays for paints and glues offer more control; cleaning materials do not need aerosol sprays. Phase out your use of aerosols in the office and you will probably end up saving money. Remember, those aerosols marked ozone-friendly have removed CFCs, but have added other atmospheric pollutants.

Provide disposal facilities for sanitary napkins

The 1936 Public Health Act forbids the disposal of any article that might block the flow of the sewage system. Women's toilets must have facilities to dispose of sanitary napkins if they are used by more than ten employees or members of the public. There are sound environmental reasons for not flushing sanitary napkins down the toilet: after they are passed through the sewage system they end up in our seas and on the beaches. However, more than half of the US workplaces have less than 10 employees and are therefore not covered by the Public Health Act. Disposal facilities ought to be provided in your workplace as a matter of course.

Recycle your rubbish

Rubbish collected in the workplace can often be more valuable than you think. Discarded computer paper is the most highly prized of all recycled waste, and it fetches a high price. Cans, bottles, and even food wastes can be recycled or made into compost for the office garden plot.

Support your cleaner

The cleaning staff faces many hidden hazards in the workplace. Handling detergents and cleaning fluids can lead to skin problems. Solvents used for floor-stripping, window-cleaning and removing graffiti can cause skin irritation, drowsiness, or even unconsciousness. Airborne dust causes eye, throat, and nose problems. Toxic gases can be formed when bleaches and disinfectants are mixed together. These can cause lung damage. Allergies to cleaning fluids, accidents caused by wet floors, back and knee injuries and the general risk of disease from cleaning up our waste often make the cleaners' jobs the worst in any workplace. A large proportion of cleaners are migrant workers, mostly women, who are given little or no training and have few rights and liberties in their workplace. Support should include training, getting rid of toxic cleansers, and using safer, environmentally benign alternatives.

Say no to a company car

Unless your job involves deliveries, do not accept a company car as part of your pay package. It costs the taxpayer millions every year in lost income when cars are processed as belonging to the company. The pollution created by these cars amounts to millions of tons of carbon dioxide, the most widespread greenhouse gas. If employees in your company need individual transport, make sure it's at least lead-free and buy a new model that has a catalytic converter.

Share your car

If you cannot take the bus or train to work, then do not travel alone in your car. Start a carpool. This way you will be sharing the costs, as well as cutting down on both the pollution emitted and the numbers of cars on the roads by up to 75 percent.

Provide company bikes!

Why not offer a company bike instead of a company car, which guzzles fuel and pollutes the air? Bikes are much cheaper to buy: even an extremely good bike costs under $400. They have virtually no maintenance costs compared to cars, they create no pollution and they help to keep us fit.

Be kinder to your environment

Workplaces can do a lot for the environment. You can support groups locally, giving donations or goods in kind. You can set aside an area in your grounds to go green, planting trees and encouraging flowers. Have real plants in the office, not plastic ones, and encourage people to look after them. Creating a green environment in your workplace will have positive effects on staff morale and create a genuinely better working atmosphere.

Build a kestrel box

Some birds prefer the warmth of a city. Office blocks and factories can create an artificial home for the kestrel, often to the delight of workers. This magnificent bird, with a wing span of up to 32 inches, can be found in Manhattan and is a familiar sight at the sides of roads, where it looks for voles and insects at a height of up to 40 feet. In cities kestrels like to nest on top of tall buildings and you can help them by building a home for them. Around urban areas try looking upward more often: you may be surprised to spot this reddish-brown falcon.

Tell someone else

Despite the current "greening" of companies, some work-places will not give up some of their profits to clean up their act or protect their employees without a fight. Though legislation provides some protection, a worker who complains may risk being passed over for promotion or even sacked. When the environment or the health of the public or other workers is at risk, profits must come second and someone has to speak up. Get help from outside sources like an advocate, an advice bureau or a trade union, and prepare your case. Perhaps you should even consider sending the information to a newspaper to expose extreme practices.

Sport and Leisure

Decide where to eat

Every day 300 million people at out in restaurants in Europe and America. Where they choose is as important as what they eat. Some of them opt for fast-food outlets, munching through hamburgers and "junk food" at amazing rates. Using millions of tons of waste paper and plastic, fast-food outlets are spreading around the world, changing the normally healthy diets of the young into food fads. The London Food Commission reckons that every hamburger we eat may contain offal—the internal organs of animals, like the pancreas, stomach, lungs, and testicles—and 100 percent beef burgers can legally contain 30 percent fat. Fast food is high in sugar, salt, and fat, and low in protein and vitamins. Choose well to enjoy your entertainment, do not waste packaging and you could feel better for it, too.

Throw an organic dinner party

Have everything, from wine to dessert, organic and sugar-free.

Watch human circuses, not animals

We can abuse animals by degrading them to perform for our amusement. In circuses we watch as chimpanzees dressed up in human clothes have tea parties and jump through rings of fire on roller skates, and tigers, elephants and lions run through hoops, and have whips cracked at them to prove our power over them. These tricks are not always "amusing" to the animals concerned and stories of inhumane practices and cruelty are not uncommon. We can derive pleasure from watching humans in new-style circuses, such as the world-famous Circus Oz group from Australia, with their fantastic acrobatics. There is little justification for having animals prance around for our pleasure.

Leave motorboats for the bath

One of the most annoying forms of noise pollution is a high speed toy motorboat in a public pond. Unfortunately this does not just annoy those people who can hear it up to a mile away, but it also upsets and stresses birds and wildlife. This sort of executive toy ends up being a public and environmental nuisance.

Do not collect butterflies

Butterflies are a good indicator of the state of our environment. When they can be found in abundance they show the local ecology is generally in good form. But species have been declining around the world in recent years. Of the 700 species of butterflies and moths to be found in the United States, the average person will probably only ever see about 5 percent. Nearly 100 species are listed in the Red Data Book as seriously threatened by pesticides, a loss of habitat and climatic changes. Collectors, however, are responsible for acquiring and mounting large numbers of butterflies; it is illegal to collect those that are endangered.

Take photographs

Become an expert in your local community and on holiday. Take photographs of everything. Photos are modern-day records of how we live, and they will perhaps become poignant reminders of wildlife scenes of today. In the United States the National Audubon Society produces a diary with beautiful pictures of wildlife scenes supplied by the public.

Write it down

Much of the poetry and writing we take for granted today has been inspired by nature and by beautiful natural surroundings. D. H. Lawrence and Thomas Hardy, Dervla Murphy, and J. D. Salinger have shown us a countryside and world to be admired and loved. Hilda Murrell, the anti-nuclear campaigner, kept an inspiring countryside diary, published after her death. Write about your own area in letters, poems, journals and diaries and send your contributions to local newspapers and community and theatre groups.

Knit it

Knitting can be more than a hobby. Dramatic designs and warming woollies can cheer up just about anyone. Knitting saves money and can also be soothing—and fun! Use good quality pure wools, avoiding those mixes made with petro-chemical derivatives. Get yourself some patterns and needles and you're away!

Check the water for water sports

Blue-green algae is causing problems for those who take part in water sports. In 1989 marinas, swimming areas, and water-sport and sailing centers were closed for weeks as a thick blue-green algae spread over still-water areas in Britain, Canada and some European countries. Local people were taken to the hospital with serious illnesses as a result of swallowing the water, and at least twenty-four dogs died in the Anglian region alone. The poisoning is thought to be connected to high phosphate and nitrate levels from sewage and farming in the water, but scientists have yet to pinpoint an exact cause. Individuals are advised not to swim or take part in water sports in areas affected by this algae.

Use lead-free fishing weights

Every year in the United States some 5,479 waterfowl are poisoned by lead, mostly from fishing weights used by enthusiastic anglers. In the last few years much concern has been raised about the vast numbers of fishermen discarding lead weights into waterways around the country. Lead can be a killer. It is possible to use tungsten polymer or tin rubber polymer alternatives. If you still use lead—stop.

Watch your nets

Serious fishermen who use special nets or lobster creels for catching fish could be responsible for killing otters. The otters look in the nets for food and an unattended basket becomes a death trap.

Check your boats

If you are lucky enough to own a boat or yacht, then you will have to be especially careful about using antifouling agents like parachlorometacresol, which is highly toxic and has been found to create long-term environmental damage and pollution. Be especially vigilant with chemicals that you do not know, and use only the minimum amount necessary. Tributylin compounds have been found to be persistent pollutants, causing damage to oyster beds and dog-whelk populations.

Anglers—protect the water first

Anglers, who are often the first to notice a decline in the diversity of aquatic life (an indication that the water is poisoned), can get involved and campaign for a cleaner water. Pollution from industry kills aquatic life.

Do you need to watch the dolphins?

Popular display cetaceans are the killer whale, the rough-toothed dolphin, the pilot whale, and the bottlenose dolphins. There is widespread concern over the physical and mental stress this can cause, and the high mortality rate of captive cetaceans. Some 4,580 small cetaceans have been caught so far for display and research. They may look beautiful in a pool, but they are far happier at home in the ocean. Do not encourage the spectacle of watching these beautiful creatures perform tricks for our cheap amusement.

A-hunting we won't go

Mink hunting could be endangering the otter even further. It is not illegal to hunt an otter, but it is to kill it. Because of this the mink has become a ready substitute for hunters. They are followed by packs of hounds along river banks causing disturbances and panic not just among the mink, but among all other wildlife in the vicinity. This ''sport'' is carried out between the end of one fox-hunting season and the start of another. There is no scientific or historical reason for hunting mink or otters.

Don't hunt

Hunting animals for pleasure is not a sport. A sport usually has two willing participants, but hunting rabbits, squirrels, deer, and other animals is nothing more than a barbaric act for humans to inflict pain and cruelty over another living species. The whole of our ecosystem is being destroyed by our acts of vandalism. Last year Americans killed 4 million white tailed deer, and the population is dropping steadily. In Maryland alone 33,000 deer were killed from a population of 100,000.

Leave the ducks alone

In 1989 5.2 million ducks were shot by hunters from a known species number of 60 million. Using lead shot on ducks further causes another 1.5–3 million deaths per year from ducks ingesting lead. In the state of Louisiana for every one duck shot legally, four were killed illegally, and now the species is at an all time low. Don't shoot ducks.

Cherish the bear

Hunters tend to think of animals as trophies on their living room walls and the American hunter is no exception. Last year hunters killed 21,000 black bears legally and poached thousands more. The grizzly bear is now endangered and soon all other North American species will follow.

Look after golf courses

Golf courses are great places for wildlife, provided pesticides aren't sprayed over them. The land set aside for this sport makes a significant potential wildlife reservoir. Unimproved "rough" lands are the most prolific in wild creatures and offer a haven for rare species like the natterjack toad and great crested newt. Badgers and foxes like to breed in this terrain and the number of birds frequenting these areas increases every year. Protect the wildlife of the golf course: don't spray pesticides and chemicals. If you are a member of a club, then make sure committees and stewards know of the importance of their land.

Tanning Salons are a hazard

The American Medical Association produced warnings in 1985 against the continued use of sun lamps. In their report they concluded that "there is no known medical benefit obtained from cosmetic tanning. Exposure to high intensity UVA radiation in the tanning booths currently in vogue is a health hazard." Standards for sun lamps users are scarce, especially for those who use them at home. Most specialists advise people who care about their skin to avoid using tanning salons. A British Consumer Association report concluded that at least one in five home sun lamp users does not use the protective goggles that are essential to protect you against eye damage.

Don't support bow hunting

Hunting animals with a bow and arrow is a barbaric means of killing them. This ''sport'' involves shooting at the animal and then tracking it, a process that can take from 10 hours to 10 days. At least 50 percent of the animals escape, only to die later of wounds and infections. This sport should not be tolerated.

Reinstate the mourning dove as a symbol of peace

The mourning dove represents America's number one game species and yet internationally it is the symbol of peace and tranquillity. Some 50 million of these birds are shot purely for sport. The maximum amount of meat you can get is 1 ounce. This trusting bird could well be endangered in the near future.

On the Road

Do without the car

During 1987 there were more than 400 million cars in the world, mostly in North America and Europe. They use one-third of the world's gas and produce one-half of all the world's air pollution. The University of California estimates that the use of gasoline and diesel fuel in the United States alone may cause up to 30,000 deaths each year. Air pollution is thought to cost $40 billion per year in health care and lost productivity. The car is the most dangerous, uneconomic and environmentally polluting form of transport that has ever been invented. Isn't it time you looked at the alternatives? The suggestions below should be looked at if you feel you really *need* the car, but its use cannot be endorsed.

Forget the parking—leave your car at home

Parking places near shopping centers and offices are becoming increasingly expensive. A parking space in a New York City condominium garage will set you back $29,000. In Toronto, Canada, the city authorities have encouraged high parking prices to dissuade people from using their cars in the city because of excess pollution, traffic jams, and accidents. Next time you go to town, leave your car behind and the land space now used for parking lots can be converted to more ecologically sound uses.

Take the bus!

A bus can carry over a hundred passengers for less fuel than a single car. Clearly, one bus takes up less room than the hundred cars needed to carry the same passengers if they were driving on their own. If only 5 percent of our car journeys were made by bus, we could produce 250,000 fewer tons of dangerous emissions like carbon monoxide, sulphur dioxide, nitrous oxides, and lead every year. Buses are cheaper than cars, and the more people use them, campaign for them and insist on them, the more buses there will be. So, next time, take the bus.

Use your feet

Walking is a very important part of our lives. About 74 percent of all recorded journeys are of 4 miles (8 kilometers) or less, and most of these are on foot. A walker is more likely to be killed on the road by a passing car than a car passenger is, and yet we give little opportunity for safety on the roads for pedestrians. Curb and pavement maintenance, pavement parking, dog excrement, and lack of safe crossing places are just some of the problems for the average pedestrian.

Hire a car

There may well be times when you just can't do the things you want to without a car. Why not find a good car rental service and hire a car as and when you need it? Make sure the car runs on lead-free gas and has a catalytic converter. It works out cheaper in the long run, as you don't have to pay for road tax, insurance, garage fees, etc. Be sure to look after the car, though: a car abused by many drivers lasts only a short time.

Cycle for safety, cycle for the planet

Cycling is by far the most energy-efficient form of transport ever invented. Bicycles are cheaper to manufacture than cars, easier to park, safer for pedestrians, produce no pollution, and are easy to maintain. Go by bike—there just isn't a better way to travel.

Cycle round your town

A cycling demonstration is a great way of gaining support from other cyclists and swelling the number of cyclists in your area. Get together a group of cyclists to cycle round your town. Survey potential cycle lands and lobby your local authority. What about asking other cyclists to sign a petition for better road facilities?

Let the train take the strain

The number of people using rail travel in the United States is increasing even though public subsidy is decreasing. The current government policy is to reduce the subsidy to zero in the next few years. Across the rest of Europe we see an altogether different picture. Turin in Italy comes top with a subsidy of 87 percent and Rome and Rotterdam second with 81 percent. For long-distance traveling, the train is the most efficient and cleanest form of travel. And it is often the pleasantest: you can walk around while the train is moving, travel easily and safely with children, eat and even sleep on the train. Campaign for a reinstatement of the train subsidy, get polluting cars off the motorways, and get people back in the railway carriages!

Get a ticket

A railcard, travelcard, or season ticket will encourage you to use public transportation and keep your costs down. It makes life easier for conductors and ticket collectors, and this in turn speeds up human traffic on platforms and on the bus. The income from regular season-ticket holders allows the transport system to plan ahead, which means a better service. Last but not least, you waste less paper by having only one ticket per month or year, not of two or three a day.

Share a car

You may not own a car, but you could still have access to one if you need it. How about sharing one car with several friends who live nearby? Or linking up with community centers or organizations to "loan" cars when you really need them?

Avoid driving in bad weather

Avoid driving in bad weather, not just because of the added risk of accidents, but because there is around a 3 percent fuel loss for each 10° F (6° C) drop in temperature. Driving in the rain and snow reduces gas economy by about 10 percent, and the wind reduces mileage. Try to make your journey at another time and, if this is impossible, think about taking public transport as an alternative.

Avoid using a car in cities

Driving your car in a city uses at least twice as much fuel as driving a similar distance on the highway. The average speed is often about 10 miles (16 kilometers) per hour in cities like London—which is slower than the horse and cart of a hundred years ago! With buses, taxis, bicycles, subways, and walking as alternatives, your car should be bottom of the list in a city.

Take the junk out of your car

Before you set off on a long drive, check your car. You may feel that it is vitally important to carry around your tools and a spare battery, but the lighter your car, the less fuel it will burn and the less pollution it will create. Every 100 pounds (45 kilograms) of extra weight decreases gas efficiency by 1 percent.

Buy a "greener" car

If you can't live without a car then you should shop around and buy a car that does as little damage to the environment as possible. Buy one that is able to run on lead-free gas.

Make sure it is fuel efficient, and do not buy a car that is larger than you need. This could be one of the biggest choices you make for the environment, so do it well and with all the necessary information. Get in touch with environment groups that campaign on car use and collect your information carefully.

Consider a secondhand car

A secondhand car may initially be cheaper to buy, but it could cost you much more to run. The older an engine becomes, the less efficient it is and the more fuel and oil guzzling it becomes. The life span of cars has decreased very quickly. Cars are made with built-in obsolescence in mind, so that you must buy a new one more often. Secondhand cars tend to need more work and more attention than new ones do.

Install a catalytic converter

If you have to have a car, then you should consider buying a catalytic converter, which will cut the emissions of carbon monoxide and nitrogen oxides from each car by 75 percent. The gases emitted by cars cause serious damage, such as acid rain and the greenhouse effect, which contributes to the warming of our planet. It is vital that we consider the use of catalytic converters. A British company, Johnson Matthey, is one of the leading manufacturers of catalytic converters worldwide, yet produces most of these for overseas markets.

Avoid short trips in cars

The easiest way to waste gas is to set out on a cold day for a short trip in your car. You can squander all the precious fuel you have saved by looking after the vehicle and being safety conscious. Cold weather means that you will have a longer warm-up time for the car, wasting fuel further. Walk or cycle for short journeys.

Check your brakes

The brakes on your car are a vital component and you should check them regularly. A loss of brake fluid is the commonest type of brake failure, and is caused by leaks in the hydraulic system. Check the fluid level regularly and you will save fuel and energy—as well as your life.

Look after your car

United States law states that a car must pass inspection. Most cars must be tested annually, but to save energy and cut down on pollution your car should be regularly tested and maintained anyway. A car tuning could give you a 5 percent fuel-usage improvement, and checking on brakes, reflectors, tires, seat belts, exhaust, indicators, shock absorbers and lights all ensure that your car is not a safety hazard.

Plan your trip

Plan your trip carefully in order to use the minimum amount of fuel necessary. Traveling during nonpeak hours, having to stop at the least number of traffic lights helps save on fuel. Avoid crowded cities as much as you can. Call your auto club and get help with planning the trip.

Recycle CFCs in cars

An air conditioner in a car causes an average release into the atmosphere of 2.5 pounds (1.1 kilograms) of CFCs per car each year, according to the Greenhouse Crisis Foundation. Recharging your air conditioner causes another 1 pound (0.4 kilogram) to be released. This is one of the main sources of chlorofluorocarbon emissions in the United States and, as more and more cars are fitted with air conditioners across Europe, the problem will escalate. Machines are available that can recycle and save the CFCs during the service and repair of your air conditioner. Insist that this is done.

Use the best quality oil

Using oil in your car is vital for a long-lasting, efficient engine. The better the quality, the more fuel efficient your car will be, as this reduces engine friction. You should change the oil as often as is recommended by the manufacturer. Keep a record of each change, for easy reference.

Use unleaded gas

Though levels are falling, car exhausts put out tons of lead. This contributes to serious illnesses, including brain damage in young children. Start using unleaded gas immediately if you have not already done so.

Stick to 55 miles per hour (88 kilometers per hour)

Driving at a steady 55 miles per hour on the highway or road is safer and can reduce the amount of fuel you use. Driving at 70 m.p.h. will be as much as 25 percent less efficient in fuel consumption, and thus cause a lot more pollution. Speed, of course, also means that you will be more likely to have an accident.

Check your tires

Check your car tires at least once every two weeks and before any long journey. Worn or damaged tires are both illegal and dangerous. Inspect the treads and walls for flaws and sharp objects. Bulges in the sidewalls are particularly unsafe. When correctly inflated, car tires can save up to 10 percent on performance efficiency.

Use radial tires

Radial tires can improve your fuel efficiency by as much as 4 percent when compared to cross-ply tires.

Use tinted glass in your car windows

According to the government in the United States, new studies show that light colored cars with tinted glass need less air conditioning than other cars. Avoid buying a car with air conditioning if you can; they all use CFCs, which damage the ozone layer. An air conditioner uses more fuel per mile as well, costing more money and causing more pollution. Why not ask about solar-powered air conditioning, now widely available in the United States?

Learn to drive better

A driving instructor I knew once said: "If only people would come to me and ask me to teach them how to drive safely for the rest of their lives. Instead, they ask me to get them through their driving test." Our attitude to driving is something that does not only save lives. Slowing down well in advance of traffic, accelerating gradually and driving at a reasonable speed saves gas and creates less pollution, too.

Avoid the "extras" when you buy a car

These days the huge market for cars means that we are given an ever-increasing number of "extras" to entice us into buying a particular model. Electronic windows, gadgets and devices for every conceivable whim, from lighting a cigarette to turning over your cassettes. All of these, plus others such as power steering, power brakes, and air conditioning add weight to the vehicle and increase the amount of engine power needed to run it.

Warm up without air conditioning

The biggest users of air conditioning are cars, in which CFC-12 is used—which is an ozone depleter. The CFCs can be released when the system leaks or in the fitting of the air-conditioning unit, and the mixtures of CFC-12 and DME azeotrope, which is being presented as a substitute for CFCs, still cause problems, as they are greenhouse gases. Do not buy a new car with air conditioning.

Recycle your car tires

Recycling schemes for car tires save energy and resources but are underutilized. When you recycle 2.2 pounds (1 kilogram) of rubber from a tire you can save up to 71 percent of the energy it takes to produce rubber from plantations. In the United States, where some 657,534 million tires are produced every day, only 197,260 are recycled per annum. Companies are starting to introduce recycling schemes, and most tires now contain about 10 percent recycled rubber. Ask when you next get your tires changed.

Keep off the pavement

Some car drivers park on the pavement because it stops their cars from being hit, especially on narrow roads and sometimes because of sheer laziness. But it is a hazard, causing millions of pounds worth of damage to curbs and walkways not originally built for tons of metal. It also causes havoc for pedestrians, especially the blind and disabled, and those with children in strollers. Park sensibly, and walk the extra few minutes if necessary.

Don't support new road-building schemes

In California, the Department of Transportation admitted that after spending $61 billion on new roads, no further road-building would ensure free movement for drivers. Successive governments have thought they could solve the crisis by building more roads; they still have not learned better. Study after study has proved that all new road-building schemes help for a short time only, but then they attract more cars, vans and heavy trucks, and simply aggravate congestion and create even more pollution. Do not support new road schemes, do not drive unless absolutely necessary, especially in cities, and you will stop the madness of the new roads.

Allow "traffic calming"

Six hundred and fifty Americans are seriously injured on roads in built-up areas each year. Cars have been regarded as more important than pedestrians for a long time, making roads and cars the main feature of our communities. Pedestrians lose out. Pollution, noise, and lack of safety are just some of the problems. Many roads in residential areas could be closed off to cars. "Traffic-calming" schemes include road humps, creating pedestrian areas, building wider roads to allow one-way traffic only and using bricks and cobblestones to create a different road surface (which makes cars slow down). All these measures can be carried out by local councils and officials, but they must be agreed to by residents. You could put these ideas forward and argue for a better environment for your area.

Don't drop litter from your car

There's nothing worse for a cyclist or pedestrian (or even another car driver) to see than a driver winding down a window and throwing out cigarette butts or candy wrappers. It's not simply a bad habit. It's an environmental hazard. Don't do it.

Don't salt the driveway

In freezing weather ice and snow can be dangerous, but using salt to clear your driveway and country or city roads can have environmental effects. Nearby water becomes salinated and can harm wildlife. It also damages your car. In very cold countries salt is sometimes necessary, but grit is usually more than adequate.

Paper airplanes

We use staggering amounts of paper in airplanes—as napkins, as head rests, as wrappers for food and utensils. All of it is, of course, bleached white and made from trees. Next time you fly, try saving as much paper as you can—don't take it in the first place. I find that if I take the paper and try to give it back afterward it is always just thrown away. Every minute of every day there are 1 million people in the air. This means that billions of tons of paper are being used and thrown away, or not used and thrown away!—literally millions of trees are unnecessarily destroyed.

On Vacation

Don't book a package vacation

In the usual course of events, package-tour operators have little or no sensitivity for local environments or communities so long as they are clean and vacationers enjoy themselves. The local communities are then left to deal with the rubbish and waste left behind, while the often badly needed foreign exchange goes to the tour operators. Do yourself and the country you want to visit a favor, and do not book a package vacation.

Invest in development vacations

Some travel agents act as charities and kick back all of their profits into supporting good, sustainable development projects. One particularly good agency is North–South Travel, which promotes vacations between people in the north (Europe, North America, etc.) and people in the south (India, South Asia, etc.). By booking your vacation through them you could be supporting worthwhile projects in the country you visit as well.

Support vacation firms that care about the environment

There is a growing number of vacation companies that kick back profits and resources into the environment. Some operate in the country where the vacation is to take place; for example they support wildlife sanctuaries and conservation groups. Some help fund key research projects that study the effects of man on the environment. You can support these companies by booking your vacation through them, and asking your regular agent to consider similar kinds of support.

Choose an unusual vacation for adventure

Did you know that there are 92,000 North Americans on a waiting list to go to the moon? Most of us expect some interest and enjoyment from our precious vacations but the thought of being shot into space is definitely not for me. Choosing vacations that will have little impact on the environment (traveling to the moon fails miserably on this score), means proper planning and organization beforehand. Hiking and camping trips are great as long as you are not planning to tramp through an ancient woodland growing orchids. Vacations planned with sensitivity take into account the environment rather than intrude on them. Adventure vacations are best when you leave the area intact for others to enjoy.

Take an activity break

Rock-climbing, horseback riding, hill walking, sailing, canoeing and mountaineering—all are invigorating experiences for those who prefer an activity vacation. There are hundreds of choices and most seek to instill a healthy respect for the environment as an important part of the break—and there's always space for beginners!

Go on a pick-your-own farm vacation

There are still a number of farms with "pick-your-own" facilities, and they often have a wide variety of fruit and vegetables. You can mix the pleasure of a vacation with the pleasure of choosing your own fresh food. Of course, this is only good as a vacation when the particular crop you are interested in is in season. Perhaps everyone should have a two-week vacation at harvest time, in order to pick, bottle, and preserve the best fruits and vegetables in season for the coming winter months.

Don't choose hunting vacations

Hunting vacations provide a most barbaric form of pleasure, especially when capturing or killing endangered species is the most important activity. It is estimated that in the United States 250 million wild animals are killed by people on hunting vacations every year. One of the most prized trophies for hunting expeditions in South America and Africa is the mountain gorilla, found in rain forests (which are also under threat), and now on the verge of extinction. Rhinos, leopards, tigers and reptiles are also hunted. With so many of these species seemingly heading for extinction, usually due to human activity, we should not support exotic blood sports.

Don't watch or support bullfights

Spanish, Portuguese and South American bullfights are controversial and bloody ways of celebrating life and death. Originating in Crete thousands of years ago, when they were seen as offerings to the gods, bullfights have now become simply a tourist sport. Every year many thousands of bulls and hundreds of horses are killed and maimed. Over $100 million is spent each year in tickets for Spanish bullfights, much of this money coming from tourists. When you go on vacation to countries where this activity is practiced, do not support it.

Visit a national park

The vast national parks of Canada and the United States are often organized for the tourist, with campsites, water taps, maps and scenic viewpoints. Some are left completely wild, and great care has to be taken before entering them. They often have fabulous wildlife and spectacular views. The national parks of the British Isles are also worth a visit, and offer hotels, vacation centers and bed-and-breakfast accommodation for short or long breaks.

On safari, don't run the animals into the ground

The delicate balance of water in the body of gazelles means that they can survive for long periods without adequate food and water in hot climates like Africa. When you come along in your jeep wanting to take photographs of these beautiful creatures you can inadvertently kill them. If you chase the animals to get your picture they exhaust themselves and often die on the spot. Safari vacations are questionable on these grounds alone, and they need to be carefully managed to make sure that the vacation is not just another way of exploiting animals in the wild.

Conserve the skis

Skiing vacations have become another way for vacationers to become aware of the environment. Over the past few years, lack of snow on European slopes has meant that the US market has grown. The change might be due to global warming, but no one is quite sure just yet. It must be remembered that the skiing itself threatens and disrupts the ecological balance of the local area, however. Bird habitats have been depleted, deforestation across the mountainous Alp regions has further threatened wildlife and landslides have been more frequent because of lack of tree cover.

Canoe

Canoeing vacations are not for the faint-hearted. They can be strenuous, but are exciting, and highly recommended. I was lucky enough to spend some time canoeing in Canada recently, and I saw more wildlife from the water than I had ever seen before. Canoes don't harm the environment—they've been used for hundreds of years by native Indians—and they are relatively safe and cheap. If you go camping and canoeing, be sure not to leave your trash around. Take it home with you and dispose of it carefully.

Have a cycling vacation

Not as strenuous as it sounds, a cycling vacation can consist of anything from day trips to long hauls. No pollution, no gas, sheer person-power to get you around. Many vacation companies have cycling tours as options, but you can plan one yourself quite easily, with good maps and guide books.

Study while having fun

Try choosing a "green" study vacation. Field-study groups, summer camps, and courses offer a practical way to spend your free time and learn in the process. The idea of learning while having fun might seem strange to some people, but it can be more fun than you think. Environmental awareness, ecology, and nature-study classes could be a wonderful way of discovering all the things you missed at school. You would meet like-minded people, too.

Go remote

Remote cottages and cabins in wilderness areas offer great vacations for those who want to get away from it all. Nature and wildlife magazines often carry advertisements and you can walk, birdwatch, and relax.

Go gourmet

International vegetarian handbooks can suggest ways of eating well while on vacation. Vegetarian meals can be found in almost every corner of the earth, so try alternatives.

Work on an organic farm

Working weekends on organic farms offer vacations with a difference. You can work your way to health! Spend time learning about organic farming, eating great food, and gen-

erally being useful. It can be a positive break for city dwellers.

Go walking

Forget the car! Go on a walking vacation. All across the world,
walking vacations are becoming more popular. You can choose
between mountains, hills, valleys and low-lying areas with
footpaths. Walking saves energy, increases and stimulates
good health, and you can see much more from the road.

Go birdwatching

Birdwatching is one of the safest ways of admiring these beautiful creatures. Vacation specialists provide every possible
comfort, and you can monitor birds for wildlife groups, too,
doing an important service. Too many birdwatchers can destroy the habitats of the very birds they have come to see, but
most organized vacations are exceptionally careful.

Have a green Christmas break

At least one hotel in Britain offered a special green Christmas
break during 1989/90, and after its success, there will no doubt
be more in the coming years. A five-day break included vegetarian food, a visit to a nature reserve (as an alternative to the
animal-hunting holidays of days gone by) and even a Santa in
a green costume with fake fur!

Visit a city farm

City farms are springing up everywhere. They offer fantastic
facilities for education and study groups, and are also great for
day trips and interested armchair conservationists.

Don't take the roof rack unless you need it

An unloaded roof rack can increase your car's gas consumption
by nearly 10 percent. If you need to use one for your vacation,
then lay the largest suitcases flat, with smaller items packed at

the back on top of each other. The load at the back should be higher than the front to get the best speed and gas consumption.

Use organic insect repellents

When you go to a country with mosquitoes, midges, or black-fly, oil of citronella is a good natural insect repellent. Nonorganic commercial repellents contain diethyl toluamide (DEET), which is immediately poisonous if swallowed and has been responsible for many deaths. It is also absorbed through the skin when it is rubbed in, and tests show that repeated applications of a quite gentle solution can lead to brain disorders and effects on the central nervous system. Never use it over a long period. Unfortunately, the hotter you get, the more likely it is to be absorbed through your skin. I wouldn't recommend using DEET on children at all. A solution of vinegar can create a bad smell that wards off most insects. There are some new, commercially available, organic repellents, made with mixtures of essential oils that not only keep your skin supple but also create a nasty environment for insects. Where necessary, don't forget to take your malaria tablets!

Look out for a safer sunscreen

If you must sunbathe, then use a sunscreen. There are increasing numbers of lotions and oils available containing aloe vera, PABA and cocoa butter, which are pleasant and comfortable to use. They are also often made now without animal-testing. But the only real way to protect yourself from harmful ultraviolet rays from the sun (which are now increasing in strength because of the thinning of the ozone layer) is not to sunbathe at all.

Make your time away a no-sunbathing vacation

The public is still generally unaware of the serious risks posed by sunbathing. 8,800 people are expected to die from skin cancer in the United States in 1990. People most at risk are those who have a short, sharp exposure to the sun in intensive sunbathing sessions while on vacation. The ozone layer, over 15 miles (24 kilometers) above the earth, is thinning because of our use of CFCs and halons. The incidence of skin cancers will increase as the ozone layer becomes less and less able to screen out the sun's harmful ultraviolet rays. In countries such as Australia, where the number of skin-cancer deaths is enormous, government bodies are advising the general public to stay out of the sun.

Reject excessive packaging

You are more likely to come across restaurants that serve individually wrapped portions of butter, sugar, salt, pepper and ketchup while in airplanes, airports, and vacation areas. Some countries insist that restaurants use these individual packages to cut down on possible food contamination, but there has been no clear evidence to show that contamination of foodstuffs is higher when you share a ketchup bottle with someone else, provided the restaurant is clean in the first place. Don't take condiments like sugar and salt unless you know you are going to use them. If you reject them immediately they are likely to be reused instead of thrown away.

Beware "photos ready in an hour"

The hour photo-processing booths that operate in train stations and busy shopping precincts may not be environmentally friendly. They often don't have the facilities to recycle the chemicals used to develop the films. As a result, toxic chemicals such as silver, ferrocyanide, and nitrates are added to the general waste stream. Next time you take a film to be developed, ask if the chemicals will be recycled. Use other facilities if the answer is no.

Camp CFC-free

Rigid and flexible foams, which are ozone destroyers, reach further than the supermarket shelves; they can end up in your camping gear, too. Foam mattresses use chlorinated compounds that attack the ozone layer. Get yourself an inflatable-type mattress instead.

Camp on natural fibers

Synthetic fibers in your sleeping bags aren't too comfortable to sleep in if it is hot out. The fibers don't absorb moisture very well, so you feel sticky and hot, and there is as well potential for an allergic reaction. The fibers are made from petrochemicals, which are heavy polluters, and you might be better off with a cotton sleeping bag, or a cotton sheet next to your skin if you have already got a nylon sleeping bag.

Duty free or a real duty?

Most people buy duty-free goods when they travel abroad, mostly tobacco, alcohol, and perfume. Duty-free goods are often the most wasteful in packaging of all the ways that we can buy these goods. They have boxes and layers of wrapping paper inside, and plastic coverings as well as expensive gimmicks to sell them. If you are buying duty-free, don't buy all the wrapping, too. Complain, hand the wrappings back before you leave, and send the inside packaging later. Ask the shops to cut down on the paper and plastic they use.

Take homeopathic remedies on vacation

Special homeopathic remedies are available for stomach upsets and diarrhea—very important when you are on vacation. Don't waste time trying to block and unblock your system with extra chemicals. Use arsenicum album for food poisoning and vomiting or nux vomica for diarrhea and upsets after too much rich food. Both are available from health-food stores. Better still, eat wisely and moderately.

Don't take throwaways

It has become the fashion these days to take disposable wipes with us on vacations and days out. These are often drenched in perfumes and chemicals (which make my face spring out in blotches). They are also made from rayon, the origin of which is trees, and the manufacture causes much pollution: to delignify or break down the wood fibers, chlorine is used, which results in pollution from over 1,000 chemicals, including dioxins. I always take a cotton handkerchief—it is so easy to rinse it out!

Never dump plastics

Plastics are not biodegradable. They stay around in the environment forever, and one of the biggest threats to wildlife is the casual dumping of bags and other debris that can be eaten by birds and animals. Over 2 million seabirds and 100,000 sea mammals die every year from discarded plastics. Baby seals have been found strangled by plastic bags and discarded fishnets; seabirds get caught in plastic six-pack holders, which are said to take 500 years to biodegrade; even whales have been washed ashore with up to fifty plastic bags inside their guts. The message is clear: don't dump plastics in the seas or on the road.

Eat local food

When you are away from home, do not rush to fast-food shops for dinner. Eat the food of the country you are in, and save resources. You will eat better, and more cheaply, too.

Stop the trade in endangered species

On exotic package vacations in the Seychelles, East Africa and South America, North American and European tourists buy elephants' feet for wastepaper baskets; cowrie shells; tortoise shells; coral and ivory trinkets. This souvenir trade is endangering species throughout the world. International law is difficult to enforce in all countries across the world, but it is up to you. Think carefully about the presents you bring home next time—it could be the end of the line for some species.

Refuse to have your photo taken with a chimp

Those cuddly looking chimps found in Spanish resorts for you to be photographed with are caught in the wild and treated inhumanely. In West Africa, their mothers are shot and the baby chimps grabbed by poachers. To make them seem cuddly and calm they are often injected with drugs, then dressed in clothes and foisted on unsuspecting but delighted vacationers. They also can carry diseases that may be passed on to humans, including rabies and hepatitis, and as they are smuggled into the country they are not medically examined. Protect the chimpanzee; refuse this barbaric practice if you see it and report any photographers to the local authorities or to the World Wide Fund for Nature.

Don't bring back seeds and plants

When visiting a foreign country it is tempting to bring back a cutting of a particularly special plant or flower. Don't. It is probably illegal, and, even more important, you might also unwittingly bring back all sorts of tiny insects or bugs that could destroy your garden—or even your whole neighborhood. Respect other people's countryside, and leave it where it belongs.

Protect marine turtles

Marine turtles live in the sea but return to land to lay their eggs. That is when vacationers come into contact with them. The female lays her eggs on the beach, burying them up to 35 inches (90 centimeters) in the sand. After two months the baby turtles hatch and rush to the sea. They must reach the water quickly if they are not to be eaten by dogs, birds or crabs. Humans should stay away from these hatchings, however fascinating they are.

Visit the whales—but with caution

Dolphins and killer whales have been captured and kept in aquariums for many years, giving delight to thousands of children and adults. But their average life span in captivity is only about five years; in the wild the killer whale can live to be eighty years old. If you want to see these beautiful mammals, the best and least harmful way is to visit them in their natural habitat. Visits can be arranged in California, Alaska, and Norway. But note: the sheer volume of boat traffic from vacationers can cause problems for the whales as they are viewed in their breeding lagoons. If you must go, only go on boats with responsible tour operators. The most environmentally sound way to see these creatures is from viewing areas on land.

Don't walk on sand dunes

If you go to the beach this year you might see ridges of sand and grasses building up a few yards inland. These sand dunes are one of our most delicate habitats. They offer tenuous shelter for snakes and rare lizards such as the sand lizard. Natural erosion from the sea forms ponds where natterjack toads mate. Many species of plant and grasses line miles of ridges that hold thousands of amphibians. It might take up to a hundred years to form a single ridge, but it takes only one hot summer season for the public to walk or drive over it and destroy it completely. Walk only on boardwalks. If there aren't any, complain and campaign for them. Walk on established walkways and never, never drive cars over the paths.

Stop persecuting snakes!

There is an ancient myth, probably originating in the Bible, that snakes are dangerous. Snakes are not protected in the wild, and when the public see them they often kill them in fear. Stop! They are just not as dangerous as you think. Even adders aren't lethal, and they will not attack if left alone. Snakes need an open habitat in the sun in order to live, being cold-blooded creatures, and so they risk being spotted during the day. All reptiles have fixed territories and snakes are no exception. This means that when their habitat is destroyed, they don't recolonize somewhere else—they just die out. Next time you see a snake, respect it and don't kill it or disturb its habitat—it may be the last one.

The Countryside

Visit the countryside

At least 25 percent of city dwellers have never been to the countryside at all, and yet a visit offers so much in terms of fresh air, country walks, and a different view of the world. Support your countryside, and visit it as often as you can.

Observe the countryside code

The British Countryside Commission has published a code for visitors to the countryside. Perhaps it could be adopted by other countries as an international code. Green visitors should be aware of all its recommendations.

1. Enjoy the countryside and respect its life and work
2. Guard against all risks of fire
3. Fasten all gates
4. Keep your dogs under close control
5. Use gates and stiles to cross fences, hedges and walls
6. Leave livestock, crops, and machinery alone
7. Keep to the public paths across farmland
8. Take your litter home
9. Help to keep all water clean
10. Protect wildlife, plants, and trees

11. Take special care on country roads
12. Make no unnecessary noise

Observe the farmer's code

Faith Sharp of Britain's *Countryman* magazine has devised a code for farmers in the countryside. Her advice will be of interest to all green visitors to the country.

1. Respect those who come to enjoy the countryside and make them welcome
2. Guard against damage to trees and hedges caused by straw-burning
3. Make sure that gates on public paths are in good order and easy to open
4. Keep bulls out of fields that have rights of way
5. Reinstate public paths across fields immediately after plowing, and never plow headland paths
6. Don't use barbed wire where paths cross
7. Don't remove or twist around signposts, or put "Private" notices where there is a right of way
8. Take old fertilizer bags home with you
9. Don't pollute watercourses, ponds, and lakes with farm sprays or effluents
10. Don't destroy wildlife by indiscriminate plowing, hedge-clearing, draining, or pond-filling
11. Avoid damaging verges and wildlife with heavy farm machinery
12. Don't shout at walkers in your corn. They are probably on the right of way you have planted over

Never drop litter in the countryside

Litter in the countryside is a double trap. Plastic is not biodegradable, and it can also kill small mammals and birds. If you go to the countryside, please take your litter home with you when you return. Don't feed wild animals and don't leave food behind at picnic sites and camping grounds, however tempting it seems.

Don't take cars into wildlife areas

Use your car only when you need to. Try walking more than the average thirty paces from the car into the country and you might see more than you expect.

Never climb walls

Dry-stone walls are vulnerable to the weather and human influence if they are not properly maintained. They can easily crumble and fall into disrepair: they become eyesores and sheltering animals lose their homes. Climbing walls accelerates deterioration. Never take stones from walls for the garden.

Close gates and fences

A golden rule that farmers always observe is to close gates. It sounds simple enough, but if it is not done it can allow farm animals into roads, where they can reach other fields, destroy crops and cause havoc.

Don't dump your car

Cars dumped in countryside areas are monumental eyesores. They should be disposed of carefully and with thought to the environment and the local landscape. Don't dump or abandon your car, and report those who do to the local authorities.

Save our orchards

Conserving, saving, and planting orchards is not only an important part of our heritage, but maintains diversity and habitats for wildlife. Orchards are havens for wildflowers, bees, and birds. And life around a fruit orchard can be rich and varied. You can have tastings and sales of old, unusual varieties of fruit; ''apple'' teas; cookery demonstrations; competitions; cider and perry tastings; and damson-jam sales and apple road shows. Orchards also make perfect grazing for cattle, sheep, pigs, and poultry, and are valued for walking and relaxing.

Replant hedgerows

Hedgerows protect animal herds, act as stopping points for birds, mammals, and insects and are a refuge for wildlife. They also keep soil within a boundary, so that it is not blown away by the wind. Over 250 species of wildflowers have been recorded around hedgerows, but all are now sadly under great pressure. New financial incentives for farmers may help stop the loss. Contact conservation groups who can help with replanting.

Curb the spread of invasive plants

Some European plants like purple loose strife are managing to get into American wetland areas where they smother vegetation. The euphorbia species are taking over large areas in the midwest. The Kudzu, an Oriental foliage plant, has been spreading through the southeastern United States. This plant has the capacity to engulf whole trees. Foreign-introduced species can overrun local areas, so be vigilant about bringing in species that might cause damage to local ecosystems.

If you want to protect it, buy it!

Conservation groups can protect land and buildings by buying them. Successful campaigns have been won by groups like Friends of the Earth in Oxford. From a local farmer they purchased a field that was due to be developed into part of a new highway. They then divided up the land into smaller plots and sold them, so a compulsory purchase order for the land would have meant summonsing literally thousands of landowners for just one tiny field. The group managed to raise awareness about the flora and fauna found in the field as well as question the need for the highway. Organizations around the world buy land for protection and campaigning—why not support them?

Create a habitat

New habitats can be created, and this goes some way to making up for the tragic loss of meadows and wildflower and woodland copses in the past twenty years. The habitats need to be managed properly if they are to survive, and many conservation groups and nature trusts have information and experience on their side. If you know of a derelict area or a piece of land that could be planted, start some action. Don't let it go to waste.

Support our woodlands

Growing mixes of trees native to our country help support our woodlands. More species of birds are associated with woodlands than anywhere else, and they hold a rich and diverse variety of plants and animals. Cut down on your paper use and support charities and organizations that plant new trees in woodlands and protect those already standing.

Plant what's best for the wild

Planting Norway spruce and Norway maple does not help the environment much just because they are trees. Native evergreens or other species live longer and support more wildlife species. A single oak, for example, can be a home for 284 species alone—and they are less prone to diseases—whereas a fir supports itself only. Planting native hardwood trees should be a priority. Campaign against those who plant imported varieties for pulping, for tax or for other financial advantages.

Campaign against drainage

Every important wetland site is threatened by proposed drainage schemes. Some of our most important wildlife sites, especially bird sites, have been irreversibly destroyed. If you live in a farming community you could help by lobbying locally elected representatives. Everyone, in every community, should be aware that draining wetlands for peat extraction is destroying much of our ancient wildlife habitat.

Save the forests

Our own forests are in danger, not just the rain forests in Brazil and other tropical places. Support groups that replant and look after forest areas, especially local forests and woodland areas.

Conserve the meadows

There are so very few meadowlands left to conserve that we should do all we can to preserve them. Centuries ago meadowland was unrestricted, mostly commonland and grazing space. With intensive farming much of our meadowlands have been lost. We no longer leave a field fallow every four years so it can enrich itself with wildflowers, providing a good compost for the following year's crop. About 25 percent of meadow plants are grasses. The rest consists of wildflowers like the meadow clary, cuckoo flower, or marsh marigolds. Meadow plants thrive in poor soil and grazing animals on fallow fields helped create many of our ancient and beautiful meadows. You can help conserve meadows by campaigning against development on those remaining, and joining and supporting groups who work on conservation projects.

Replant free food

Blackberries, elderflowers, and soft-fruit bushes should be planted in country areas to reintroduce free food for humans and wildlife. Plant bushes and trees yourself, encourage farmers and local councils to do so.

Support wildlife sanctuaries

Wildlife sanctuaries allow a safe refuge for all wild animals and plants from hunters and collectors. Sensitive wildlife management can encourage and enhance local ecology. You can support groups setting up or managing wildlife sanctuaries by joining them or visiting.

Protect shoulders

Major and minor roads account for an incredible amount of
land. Roadside shoulders contain a mass of wildflower species,
mammals, birds, and insects. Spraying and clearing these
shoulders destroys much of the potential habitat. You can help
to protect shoulders by not throwing litter, asking local authorities to let the wildflowers grow (this would even save them
money), and asking local naturalist or wildlife groups to carry
out surveys for monitoring and protection.

Curtail military and development activities in protected areas

The development of intensive farming, upland farming, fish
farming, commercial building, and military activities are putting extra strain on the already small areas of land protected to
enhance and preserve the natural beauty of a given area. All the
main countryside and protection organizations have asked for a
curtailment of further development on these sites.

Support the small farmer

The "holistic" approach of organic agriculture encompasses
balance, health, and wholeness as its central principles. Organic food is superior in quality and taste to chemically produced food, and using and reusing resources means that nothing
is wasted. Rotation farming keeps the soil in good condition,
and with the right amount of nutrients the organic farmer will
show a higher yield over time. Intensive farming on a large
scale means less work for farmers. This increases the number
of unemployed, brings more pollution from pesticides, and
creates a demand for high energy. Small farms employ more
person-power and less chemical and mechanical help. You can
support small farms and organic farming by buying organically
produced food.

Campaign against crop spraying

Friends of the Earth have recorded thousands of accidents that involve farmers who spray large areas of crops. Many members of the public have been poisoned. Cases have included pregnant women losing babies and small children becoming suddenly ill.

Support the use of straw

Straw-burning causes serious environmental problems—not just the danger of air pollution and smoke, but also the lost use of the straw as a resource in itself. Part of the reason it is burned is that it is too cheap to merit transportation, but animals that used to be bedded on straw are now housed on concrete slats. And straw and animal litter mix which used to be used as compost has now been swapped for a peat mix. This is increasing the rate of destruction of our peat boglands. But mixing animal waste and even domestic sewage with straw means that an inexpensive alternative is available to peat. Buying straw-based papers has also increased the market for straw. Buying only humanely reared meat products would encourage farmers to stop straw-burning and start seeing straw as a useful resource again.

Take the land out of agriculture

Land in the countryside does not need to be used continuously for growing crops. It is beneficial for the environment to increase the amount of land that is set aside for wildlife to flourish, for recreational facilities, and for conservation projects like nature reserves. Less intensive farming and less intensive livestock management helps. This means that we should have to eat lower down the food table, instead of relying so much on meat and meat-based products. The massive food surplus found in Britain can be reversed and land could be left to allow flora and fauna to flourish.

Ban aerial spraying

Aerial spraying of crops with pesticides must be the most wasteful and ecologically damaging of all intensive agricultural methods. Several surveys show that only 5 percent of the chemicals reached the intended pests. Small farmers do not generally need to use planes for crop spraying, but the larger, commercially oriented farms cut down the hedges and make huge strips of land which are then ready for bigger machines and more mechanized forms of farming. Friends of the Earth have monitored human and animal accidents due to haphazard pesticide spraying and have recommended a ban. If you live in the country, you should report all incidences of crop spraying to the local authorities and your Environmental Health Officer as a public nuisance. Protect yourself by never getting in the way of sprays and report to a doctor if you should be sprayed accidentally. Eat organically produced food.

Don't confine pregnant animals

Inhumane practices in intensive farming mean that we rear pregnant animals in cruel conditions. Sows can be kept in tiny stalls, unable to move or lie properly. This practice should not continue, and ending it is unlikely to harm the farmer one bit. Write to your supermarket and ask if it condones this sort of practice. Don't buy from companies that appear to support such cruelty.

Cull deer by shooting, not stag-hunting

Deer are hunted by experienced hunters, chased to exhaustion, and then shot. Female hinds may well be pregnant, or have calves with them. Teeth, feet, and antlers are often sold as souvenirs. The stag-hunters hunt the strong deer, not the weak, old, or lame animals—so they can't ever use deer control or culling as justification. Years ago wild predators such as wolves would have selected and killed deer, as part of the natural ecosystem. Culling deer by shooting is a more efficient and less cruel way to keep numbers down in the countryside.

Make snares illegal

Approximately one-third of the animals caught in snares are actually targeted species. Other animals mistakenly trapped include cats, dogs, badgers, deer, hedgehogs, and partridges. It is illegal to set snares for some animals, like deer. The ASPCA believes that as snares do not discriminate between species this is impossible to maintain, and it wants to make all snares illegal. Support groups that are campaigning to get all snares banned, and don't set them yourself.

Build small animal crossings on cattle grids

Many small animals find it impossible to get out of the steep pits that make up cattle or sheep grids. These grids are death traps. A small ramp or slope in one corner of the pit would enable them to walk to freedom.

Bring back the red squirrel

Lowland Britain has lost its population of red squirrels, not because the gray squirrels have pushed it out, as is commonly thought, but because of the massive loss of woodland tree cover. This is now only between 7 and 9 percent of the total land area, the lowest in Europe. The infectious diseases mange and coccidiosis are also thought to be partly responsible. Unlike gray squirrels, red squirrels have not been able to adapt to urban areas. By planting more woodland in country areas we may help them flourish.

Protect the owl

The most common bird of prey is the owl, which feeds on a diet of mice, rats, and voles. Every single type of owl is now under threat in the countryside. Our use of pesticides and chemicals on farms and our poisoning of their food has helped in the massive decline in the numbers of these nocturnal creatures that have perfect hearing and silent flight. We still know very little about them. Only 135 species exist, and most of them are now rare, so do all you can to protect them. Monitor them, report them (dead or alive), put up nesting boxes and allow rodents to breed without chemical poisoning in the countryside.

Don't dig up wild plants

Increasing pressure from industrial and housing developments, agriculture, and forestry threatens natural habitats. Plant species are threatened with extinction. The public can also play a role in saving them. On country walks many people might feel tempted to pick wildflowers or to dig up bulbs to take home to their gardens. Please leave them for everyone to enjoy.

Protect the orchid

Fourteen of Britain's forty orchid species are thought to be threatened almost to extinction in the countryside, and the rest are becoming increasingly rare because of their special and laborious flowering. It can take up to fifteen years for an orchid to flower after germination, and destruction of down land, changes in woodland management, and the drainage of wetlands have all led to the loss of orchids in the wild. The theft of wild orchid seeds is a serious conservation problem.

Protect birds' eggs

Birds' eggs found in the wild can often be valuable. But never touch eggs if you find them. If you think they are particularly rare, contact the local birdwatchers' association, who can monitor them safely.

Protect the otter

The otter must be one of America's favorite water mammals, and yet its life is threatened by our pollution. It may already be too late to save the otter, as the spread of intensive farming has destroyed their lowland habitat. Support groups who designate otter havens. If you are in a position to do so, then plant thickets, willows, and other undergrowth along streams and water banks to give otters the chance to hide in the relative safety of the banks where they once used to thrive.

Keep to footpaths

Several thousands of miles of footpaths in the United States, Britain, and Europe are favorite destinations for wildlife enthusiasts, ramblers, and hikers. Keep to the footpaths when you walk across farmland and do not spoil the crops or frighten the animals.

Protect your footpaths

An environmental disaster is looming for hill and dale walkers, who use the same footpaths so often. If you are on a footpath with muddy places, go through the mud rather than trying to edge around it, which only creates a wider path.

Campaign for more footpaths

More footpaths would take the pressure off some of the more popular paths and would add new scenic and adventurous walks for hikers and ramblers. Ask your nearest national park to create more footpaths.

Coppice and pollard in woodlands

Coppicing and pollarding bushes and small trees is an ancient practice. It ensures that some are left enough space to grow to maturity and others, especially trees, to produce a secondary layer of light and shade. This is great for wildlife, and provides extra wood for fuel, fencing and a wide variety of other purposes.

Support your village shop

Village shops and post offices are lifelines for local communities, particularly for old people. Many shops find it difficult to survive, however, when local residents go whizzing off in their new cars to buy supplies from supermarkets often many miles away. The supermarket goods may be cheaper, but people usually forget to add in the cost of traveling to the supermarket and parking the car, as well as the extra pollution caused and the time wasted in traveling. Support local village shops wherever you can; they are good news for the environment.

Use the river network

River networks around the world offer much loved forms of transport: barges in Holland, gondolas in Italy, and junks in China all conjure up romantic and beautiful images. On a practical level the river network offers a mode of transport that could have repercussions for our future transportation plans. Imagine transporting goods along the rivers on barges instead of in heavy trucks or increasing the number of people who could use the Thames to get to work. If you can, why not use the river network and take pressure off the roads?

Put up signs

Farmers, landowners, conservation groups, and local councils should work together to ensure the protection of our countryside. Signs for the public can make all the difference between knowledge and ignorance. Instead of "Beware!" why not try "Stop! Save our natural habitats from extinction"? Protection of plant and animal species might be easier for the public with a more positive approach.

THE GREEN CITIZEN

Local Community

Work with animals

Make a positive commitment to wildlife and the environment by joining the thousands of people working to save, protect and look after animals. Kennel or stable work, farming, and veterinary surgery are just some of the worthwhile careers available.

Plant a tree

We do know that the planting of trees is generally beneficial. Trees enhance their surroundings, provide a home and shade for wildlife and birds, and soak up carbon dioxide. But you should always think carefully before planting. Almost 50 percent of trees planted die within a few years because they are not chosen carefully enough for the type of soil or they are not looked after properly. By looking after existing trees and encouraging natural germination you may reduce the need to plant in country areas. By choosing good varieties with care, trees planted can survive.

Never write off trees

A tree can often survive with only one-quarter of its roots left. Dead wood even adds beauty to an area and protection for wildlife and recycles nutrients back into the soil. If you have the choice, it's better to leave dead trees where they are.

Start a tree register

Be a detective! Record the trees in your local parish, in roads, courtyards, orchards, hedgerows, and churches. You can trace the origins of certain fruit trees. Identify those trees that are local landmarks or mark boundaries.

Create a milestone

Encourage a whole new generation of environmental art in your community. Stimulate the creation of small-scale art works to express your sense of love of the natural world and its history. You can create new landmarks and new beginnings. Be proud of your place in the community.

Use the bottle bank

In countries like Holland, 62 percent of glass is recycled due to the efficient bottle-bank system. In Toronto, Canada, glass is now picked up every week at the doorstep. Until we have enforceable laws governing collections of glass for recycling it is better for the environment to take bottles, rinsed clean, to your nearest bottle bank. We need one bottle bank for every 2,000 people for effective recycling, so campaign within your local council to get this enforced.

Organize a community map

Local people, or even a commissioned artist, can draw or sew the things that they value and respect on an illustrated map. City councils, schools, women's institutes, local history, or environment groups can join together to chart and map the things that are valued in each community.

Protect trees

Cutting down a tree should be a last resort. Their importance as spiritual, cultural, and emotional companions is only just being rediscovered. Many trees are protected by law already, and it is illegal to fell them; more protection orders should be granted. Before we cut down a single tree we should consider the wildlife that depends on it, whether the timber is to be put to a worthwhile use, and whether other trees will be allowed to grow and replace it.

Don't always burn your autumn leaves

Leaves on the streets can be dangerous and slippery, but the efficiency of local authorities who clear away leaves in urban areas can upset the ecological balance of insects like wasps, moths and flies. The recent epidemic of the scale insect on urban sycamore and horse chestnut is thought to be due to the hurried burning of its predator, the chalcid wasp, which stays fixed to the underside of fallen dead leaves during autumn. Over 400 types of leaf-mining moths live on the underside of dead leaves alone and the caterpillar, greenfly, and hopperfly all pupate in the leaf and meet an early death if burnt. A better alternative is to compost the leaves, thereby keeping the streets clear while allowing natural biological control and making a rich organic mix for use on parks and gardens.

Leave some dog-free areas

Not everyone likes dogs; they can upset children and wildlife and one of the biggest environmental problems identified by the public in opinion polls is dog feces. By declaring some park and wildlife areas "dog-free" you will be enhancing the community and environment. The parasites in dog feces can cause serious illnesses, even blindness, in young children who touch it, so do not allow dogs to defecate in children's areas.

Encourage Buddleia for butterflies on wasteland

The *Buddleia davidii* is a common shrub that grows in waste sites. It can grow easily in very poor soil and the carbon dioxide from its roots forms an acidic solution to dissolve spaces in walls for it to establish itself. This plant is particularly attractive to butterflies.

Complain about the state of your pavement

Bad footpaths cause an astonishing number of accidents; at least 200 people die each year walking on the road and pavements. Regular walking in your local area will help you notice what is going on around you. Complain to your local council about the state of paths and pedestrian walkways. This will keep it on its toes and contribute to safety.

Use your motorcycle with consideration

An annoying form of environmental pollution is noise. Noisy motorbikes outside your window at 1 A.M. can be very aggravating. A noisy motorcycle is said to be able to wake thousands of people late at night so if you use one, be considerate. Getting the law changed to tighten exhaust/silencer controls would also help. If someone has removed or altered a motorcycle silencer, you can prosecute through your local environmental health officer, who will need to test the offending bike. Many motorcycles can run on lead-free gas without conversion—ask your dealer.

Ban heavy trucks

Heavy trucks are a menace. They cause damage to the environment, break up road surfaces, cause heavy wear on bridges, and cause enormous damage to the underground cables and pipes in cities like New York. Combined rail and road options are a more environmentally friendly and cost-effective way of transporting freight. By banning heavy trucks in your area you will be able to force the government and large companies to use the rail system to transport goods, reducing pollution and havoc.

Don't support incineration of garbage

Some people think that the incineration of garbage, to obtain extra energy and to dispose of our trash mountains, is good for the environment and is a new energy source. But evidence is mounting that the incineration of mixed garbage is not only dangerous but economically unviable. Large waste incinerators release dangerous gases into the air, including dioxins, and certain mixtures of waste create hydrochloric acid, cadmium, lead, mercury, and selenium. Yet if we really recycled to our maximum potential we would only have 4 percent of the garbage that we have now to be disposed of, which would be economically unviable to incinerate. Municipal incinerators are a major source of dangerous dioxins in our bodies.

Investigate weed killers in your street

If you see local authority workers spraying grass shoulders and streets with weed killers, you have every right to ask them what chemicals they are using and why. A growing number of local councils are stopping the use of chemical weed killers like those containing atrazine and simazine. Evidence suggests that these chemicals are endangering wildlife and contaminating drinking water. Laboratory evidence suggests that the chemicals are irritants and could cause severe allergic reactions. Atrazine can damage various organs and is a suspected carcinogen. Workers are increasingly demanding better safety checks on such herbicides, and some councils have agreed to use them only as a last resort.

Watch out for sea dumping

The sea is often thought of as the perfect place to dump unwanted wastes like sewage and toxic chemicals. Pesticides have been burned at sea for twenty years. Toxic wastes can be simply thrown overboard in drums. The risks to aquatic life are considerable. It is now estimated that over 100,000 different chemicals are in the North Sea alone. Drums of toxic waste that are illegally thrown overboard can be washed up onto beaches, and leaking drums cause serious pollution. If you spot suspicious drums on local beaches, report them to your local water authority immediately.

Take your hands off city walls

Graffiti makes your local environment look a mess. It is illegal and dangerous. Most so-called artists use aerosol cans, which not only damage the ozone layer or contribute to global warming but also eject tiny particles of toxic paint substances, including glues, which can be addictive and hazardous. It costs local government and companies literally millions to remove graffiti from walls, trains, buses, and stores.

Join a group

There are millions of people around the world campaigning to save the planet. They come from all walks of life, and you could be one of them. Read the newspapers, get hold of books and magazines, join an environmental group—or form one—and support those who are spending time doing research and campaigning about some of the important issues raised in this guide.

Don't drop litter

Litter costs your community money and destroys your local environment. If you decide to drop rubbish all around your house, you can continue to do so in privacy, but you should never do so in public. Get into the habit of telling people that you find dropping litter unacceptable. If enough people complained, it certainly might discourage one or two litter louts. Don't forget that wildlife also suffers when you drop litter. Birds can choke on plastic and small mammals can get very sick and even die because of rubbish left around when they are hungry. Litter also attracts and encourages rats and mice.

Support candidates who run on environmental platforms

Vote for those people who you know understand, and will carry out, good environmental policies. Ask them questions about what they will do for you in office, get commitments, and get them to stick to them if they are elected.

Write letters to elected representatives

Whether you voted for them or not, local and national politicians are responsible to you, the public. If you write to them protesting about a particular matter they will have to take note. The size of the mailbag really does influence elected officials.

Report illegal dumping

Be a super-snoop! Illegal dumping is on the increase, by industry as well as individuals. Waste matter can be extremely dangerous, both to children and wildlife, and the proliferation of unmanned dump sites, often in residential areas, is an eyesore. If you see dumpers, don't hesitate to call your local authority or police.

Complain about litter bins

If your local shops, take-away restaurants and council aren't providing litter bins, you should complain until they do. Litter with nowhere to go ends up on the pavement.

Do a wildlife and buildings survey

Examine the types of building, landscape, and animal habitat in your area. This will give you scope and inspiration for action to protect or enhance them.

Get other groups involved

Looking after the environment is not just for conservation and nature groups. Ask the local guides and scouts to help, or the local community center, senior citizens' club or citizens' action committee. Getting other people involved spreads the message and the workload.

Look after your playing fields

The long stretches of grass in playing fields offer a haven for migratory birds and seabirds. A set-aside area could be encouraged to grow wildflowers and attract butterflies and other wildlife.

Look after churchyards and cemeteries

Churchyards and cemeteries in rural areas may have escaped pesticide poisoning altogether and may be sanctuaries for all kinds of wildlife. Some cemeteries are so rich in ecological diversity that they have been turned into nature reserves. Protect them whenever you can. Record the nature and state of trees and the number of plants and shrubs. Don't let them be destroyed; make sure your local council or church group knows how precious they are.

Look after your local waterways

Improve and protect your local waterways. They are a valuable home for wildlife and all manner of plants and flowers. Local conservation groups are involved with clean-up operations and restoring such areas to their natural beauty. They can be a perfect escape in an urban area and riverbanks, waterways, and canals make your local environment green and pleasant.

Repair it

Repair, replace, repaint. Broken railings and fences, badly maintained flower beds and tumbledown walls will not improve your environment. Do something about it! Get together with people in your street or town. Ask for help from authorities and conservation groups. Once you notice what is happening you can set aside one or two days and do something really practical to help.

Get it sponsored

Activities like cleaning and clearing up your local environment can be combined with fund-raising activities for local charities and environment groups. Help them to help you. Next time you are doing something active in your local community, get it sponsored.

Look after bus shelters

Many bus shelters are eyesores unless looked after properly. Painting, cleaning up graffiti, and disposing of leaves and rubbish are some of the important tasks that need to be done regularly. Ask your local authority to help, or do it yourself.

Protect your buildings

Some local buildings are an eyesore, but some are historic and valuable monuments. Some have trees and grounds which may not have been touched for many years. Historic building societies and groups can monitor and pressurize local officials to look after old buildings. Keep notes and photographs, and offer solutions, designs, and ideas for their use.

Find a use for a building

Look out for underused buildings in your community. Could they be transformed into a home, space for a youth club, a tenants' association office or a community center? Get in touch with the owners or the local authority and start something happening.

Adopt a circle

Adopt a local piece of land, like a traffic circle or green square. Get permission to plant wildflowers, shrubs, bulbs, even trees. Agree to plan for maintenance in advance and make your area look greener.

Glorify your town square

Your town square can look great with a few improvements. Restore clocks and protect trees. Repaint seats and repair trash cans and walls. Take a photograph before and after and display them in your local library.

Start a footpath maintenance scheme

Maintain footpaths. Clear vegetation and fallen trees if necessary and get help to maintain the surface of the path.

Plan in favor of women

Women are often the last to be thought of when it comes to local environments. Planning applications should be opposed unless they consider the needs of women and women with children, as well as disabled people. Dark streets, badly maintained walkways and poor access into stores and stairs for mothers with strollers all show a local environment that is unfriendly and inconsiderate. You can have a say. Tell them what you want to see, influence your local decision-making process.

Fence with brambles and roses

Instead of barbed wire and broken glass, why not encourage brambles and wandering roses to do the job for you? They are much prettier, deter cats (and people) and will not harm birds.

Attend annual meetings

Go to the annual meetings of your local city council, civic trust, and other organizations. Find out what has been happening, and keep in touch. Even if you cannot always attend regularly, it should not be too difficult to get to an annual meeting.

Get a Dumpster

If there is a lot of rubbish in your street, such as discarded sofas, old rugs and furniture that cannot be reused, call in a Dumpster and get everyone to use it one Saturday. Get rid of unsightly and dirty garbage that attracts rats and makes your street look awful.

Use the public library

What better way to recycle books, use community facilities and read about the environment? Use your local library.

Plant flowers on your windowsill

If you do not have a garden, or you live in a flat, you can still encourage butterflies and birds and make your home look greener. Window boxes can be filled with bulbs for spring and delicate summer flowers or herbs for later in the year.

Campaign for parking spaces for cyclists

Bicycles take up almost no space at all, especially when compared with a car. They are free of pollutants and do not use precious energy. A real problem for city cyclists is the lack of safe places for parking. Why not ask your local shops, library, community center, or workplace to make sure that bike sheds, bike racks, or suitable parking spaces are provided?

Hold an exhibition

Hold an exhibition. Show past and present community lifestyles, and encourage others to join and be proud of their landscape, buildings, and heritage.

Twin your town

Twin your town and encourage community spirit and cooperation with another country. Learn to live together, and learn about how others live. Share ideas and resources.

Hold a competition

Provide the community with some fun. Hold a photography competition to spot the best wildlife in your area; or the best-kept street or avenue. Ask local businesses to donate prizes and get the local paper involved. Get people thinking about their community.

Protect your landscape

Millions of acres of land have been concreted over, stripping our landscape of woodlands, trees, meadows, and fields. Check all the planning applications you see in local papers or pinned to walls and trees. Get involved and interested in what developers are doing to your area. You have a say and you could save some valuable heritage from poor development planning.

Clean up your beaches

Join a group or contact your local authority and clean up unsightly waste from your beach or coastal area. This will not just benefit you, but also birds and marine life who live on the beaches and would probably appreciate a cleaner home.

Be vigilant

Always be vigilant. Be aware that people, authorities, and industries want to make changes all the time to planning controls, buildings, and land. Obviously not every change is bad, some could even be beneficial to the environment, but constant vigilance is necessary in your area. Keep an eye on a certain area such as the walk to the bus or subway each morning on your way to work or school. Be alert to changes.

Clean up the landfill

Landfill sites are pretty noxious—after all, they hold millions of tons of our waste—but they can offer a haven to wildlife. Over the years, as better recycling facilities are organized, we should need fewer landfill sites, but schemes can help areas that have already been filled. One successful scheme already exists in England: a landfill site has been turned into a public park. The site was covered over with 18 inches (45 centimeters) of topsoil to contain the various gases and other by-products of rotting waste. Within five years, over 145 different species of wild plants were recovered in the area.

Inspect rivers and streams for pollution

Keep your eye open for dirty water. It might be coming from local businesses and sewers that are putting chemicals into a nearby stream. Watch out for dead or dying fish and wildlife. They can be the first signs of hidden pollution. Notice which wildflowers, shrubs, and bushes are doing well. Some like acidic soil better than others, and a keen gardener will be able to spot changes in soil and water conditions. Report any changes that worry you to local authorities immediately, making sure that they are aware of your concern. It might encourage them to improve their own monitoring. Never think that someone else will do it; they won't.

Grow flowers everywhere

Flowers make a community look like a home. They add color and interest and their nectar attracts butterflies and bees. Get local garden shops to donate bulbs for hospitals and schools. Make sure local businesses have tubs or window boxes if they haven't got a green space. Let flowers cheer people up all year.

Find out who matters

Who owns you? Who do you pay money to? Who represents you in local elections? It is important that you know how your local community works. Be a sleuth, ask questions, go to public meetings, and find out who runs your town. These are often the people who need to be influenced to make positive environmental changes in your area. Write to them by name, not just by job title. Let them know that you know who they are.

Use allotments

Allotments, or small pieces of land owned and loaned by local governments, are popular all over the world. They are especially valuable for people who live in high-rise flats without gardens. Share one with a friend for mutual encouragement, though it is advisable not to take on a huge one unless you have had experience. You could grow a substantial part of your fresh food on an allotment. The organic gardener and writer John Jeavons has estimated that it is possible to grow enough fresh food for one year's consumption on a plot just 10 feet (3 meters) square.

Stand for election

If politicians ignore your environmental concerns, get them voted out of office! Local elections are easier to contest than you think, and if you feel particularly inspired, why not get involved with the decision-making process and do it from the top! Before you join a party, carry out a thorough assessment of each party's record on and commitment to the environment.

Go on trails and walks

Find out about your area. Join up with local walking clubs or set off on your own or with the family. Walking can be pleasurable, relaxing, and educational. It doesn't use any energy other than your own and makes you aware of your surroundings.

Support your local museum

Find out about your local history. It could help you in a case against bad planning applications, or when trying to save a particular green area or condemned building. Museums are a mine of information. They often hold the only records of your parish, town, or city. Ask your museum to consider keeping local records or ecological changes and even wildlife sightings. Get nature groups involved.

Save local landmarks

Look at your skyline; it tells a story. The chimneys and church spires are landmarks of your town. They represent your cultural heritage and can show the state of your environment at a glance. Is the landscape full of high-rise flats or factories with pollution billowing out? Wherever you live, look after important local landmarks. Get them listed, registered, and protected if necessary.

Lobby for safer streets

Women feel especially vulnerable in badly lit streets, dangerous-looking underpasses, and dark walkways set close to hedgerows and walls. Safer streets can be achieved more easily than you think. Good planning in favor of women and low-energy lighting make all the difference. Lobby your authority about a street near you that could be made safer. It is part of our environment, too.

Support local crafts

Local crafts are part of your heritage and supporting them benefits the environment. Local jobs are created, diversity is encouraged, transportation reduced and energy saved as local goods fulfill our needs.

Arrange a Sunday walk

People like to walk on a Sunday afternoon so why not take them on an ecological tour? Adopt a theme, such as local churches, canals, a fun walk or interesting buildings. Point out special features and encourage people to take an interest in

their area again. This consciousness-raising is an important
way to get people to appreciate and respect their local environment.

Record your history

Local residents often have much to say, especially those whose
families have been in the area for many generations. Find
out what has changed over the years and what has stayed the
same, what people like about their area and what worries them;
then use modern technology to let the rest of the world know
about it.

Hold a green fair

Environment weeks, green fairs and local celebrations bring a
community together. Share ideas and resources with like-
minded people and join up for a fun time. Get information from
ecology organizations, ask charities and voluntary groups to
help raise money by organizing bric-a-brac stalls or selling
marmalade, for example, and get quizzes, raffles, and competitions going. And have fun!

Brighten up your walls

Murals and mosaics brighten up walls in urban areas. In rural
areas, walls provide habitats for plants, insects, and all manner
of animals, including snakes and lizards. Most countryside
walls are very old. Built from sandstone, limestone, granite or
slate they act as continuous landscape features. Looking after
walls is important, especially to maintain a habitat for animals
in the face of disappearing hedgerows.

Protect buildings

Buildings can make your community look good or bad, but
they have a more important role in the environment. Old
churches can house tawny owls or pipistrelle bats. Abandoned

houses can be full of mice and bugs. Buildings over a certain age can be protected with special orders, so they cannot be destroyed without good reason. Converting houses and apartments creates useful work, saves resources and energy and provides much-needed accommodation. Protect buildings around you and make good use of them.

Don't concrete over everything

Don't allow developers to concrete over everything. Concrete seals in and kills the soil underneath it, making the land no longer productive or pretty. Opt for parks and gardens, playing fields and courtyards of fresh earth, flowers and grass.

Don't be a NIMBY

NIMBYs (Not In My Backyard) may not want nuclear waste, toxic drums, landfill sites, new factories, or blocks of apartments in their backyard, but they are quite willing to put it in yours. We all have to take responsibility for the planet. The consequences of our actions cannot be resolved by giving them to someone else. The United States ships its toxic waste to developing countries. Japan ships its nuclear waste to Britain, and Britain ships its pesticides to the Philippines. If the company you work for installs a dangerous plant in another town, it is doing the same thing. Don't be a NIMBY. Influence your community.

Ask your local media to get involved

Ask your local radio station, newspaper, or even television news to run more features about the environment. What about a regular phone-in show with local environmental group organizers on the panel? Or a program to highlight a local ecological attraction or problem? The media can get through to a lot more people and answer some of their questions easily. Try contacting your local media and asking them.

Adopt a seat

Many local communities benefit from proper seating made of ash or beech. Some have adoption schemes where you can buy a seat with your name on it in a park or woodland area for a price that includes its upkeep. This can be especially useful when local authorities cannot afford to maintain seating and let it fall into disrepair.

Take part in public meetings

Public meetings exist for you to place on record your concern or support for developments within your community. Some public meetings last only a few days and planning applications for new buildings or road-building schemes are won or lost very quickly. It is an important way for the community to say exactly what they feel about the proposal. Enquiries can be successful. Get involved; take part in public hearings regarding the future of your town.

Swap your specialist knowledge

If you are good at something like gardening, raising finance, baking or sewing, why not swap your knowledge and skills with someone who can offer you something that you need? Skills shares operate in communities around the world very successfully. You could set up a local directory and network.

Set up a local nature reserve

Nature reserves can transform a blighted area into an ecological paradise. Contact your local authority for information about the area of land you have in mind. Set up a local wildlife group and apply for grants and funding to keep the project running. Some local nature reserves have been organized with the cooperation of local schools and some with local community associations. Try to get as many different types of people involved so that everyone can benefit.

Collect stamps and bottle tops

Collect used stamps and bottle tops and send them to charities. As well as recycling, you help them to raise funds for sustainable projects.

Help feed someone when you eat

Some restaurants have joined together and supported developing groups and charities fighting against hunger in the South. Every time you eat at one of these restaurants you will support them. SOS (Save Our Strength) restaurants give excess food to the needy and run campaigns against food waste.

Order Shopping for a Better World

The Council on Economic Priorities has produced one of America's most valuable guides on shopping for ethics and the environment. At last you can tell who owns what and why. You can make an informed choice about products and companies by taking this guide with you when you shop. Products are rated according to ten major issues.

Local Government

Local politics are often easier to influence than you think. The main problem is cash, but with the right encouragement the environment can be the number one priority. Find out who your local representative is. Organize yourself to lobby and campaign to get policies like those listed below on the agenda.

Policy

Encourage local participation in environmental initiatives. Support environmental and conservation groups and other nonstatutory bodies by providing facilities and space.

Involve local people in decisions on the environment.

Provide clear, simple information to local people about their environment.

Ensure that all committees, staff and departments in local government consider the environmental impact and implications of their activities as routine.

Devise a purchasing policy which bans environmentally damaging products.

Do not buy tropical hardwoods from unsustainable sources for floors, doors and veneers, or for desks and office furniture.

Minimize the use of car-owner requirements in job advertisements. Do not ask for "owner drivers" in job advertisements unless it is absolutely necessary.

Pollution

Clean up the local area by controlling and reducing litter, particularly by providing regular collections.

Provide free Dumpsters to neighborhoods for house holders to get rid of large items with ease.

Stop municipal waste incineration.

Work with local traders and businesses to provide sponsored trash cans.

Monitor and regularly check abandoned vehicles.

Stop the transportation of nuclear waste through the area.

Find the most environmentally safe way of disposing of toxic and other nonrecyclable waste.

Regularly monitor pollution levels in the streets and provide an information service to members of the public.

Survey and replace all leaded pipes in the area still connected to the main water supply.

Enforce local bylaws on noise and noise levels, especially when relating to traffic.

Clean up eyesores that blight local areas.

Introduce a "hotline" telephone service so that local people can call about pollution worries and problems.

Encourage residents to report waste-dumping and fly-tipping and to respect their local environment.

Clean up graffiti.

Provide trash cans on council property and on council land.

Clean up any waterways.

Greening

Turn vacant sites over to local nature or conservation groups who can transform them into wildlife havens.

Improve the visual environment of local communities by creating grassed and wildlife areas, providing for better fencing and gardens and painting and cleaning up buildings.

Register and promote the use of allotments for local people and create new allotment areas.

Encourage wild species of flora and fauna whenever possible.

Encourage local businesses to open up land owned by them for public use or for wildlife.

Carry out a wildlife survey with local ecology and conservation groups.

Encourage window boxes, tree and shrub planting, and more wildflowers on council-owned property.

Provide park and street furniture that is sympathetic to the local environment, but not from tropical hardwood sources.

Agree to minimize the impact of new development plans on local habitats and wildlife areas, taking special consideration of trees.

Encourage the growth of woodland areas, especially the planting of local species.

Encourage new wildlife and nature walks within the area.

Improve the facilities offered for recreation and play, to enhance the environment and to improve opportunities.

Transport

Improve road-safety measures, including "traffic-calming" schemes and facilities for handicapped or elderly people, such as nonslip surfaces.

Repair roads, signposts, and potholes.

Provide adequate cycle-lane facilities and secure parking in all public places.

Restrict car parking in central areas and restrict trucks in specified areas.

Promote public transportation by subsidizing fares for the unemployed and disadvantaged. Provide subsidized transportation services for women, especially at night.

Always consider improving public transportation systems rather than building new roads.

Ensure all vehicles owned by officials run on unleaded gas and phase in the use of transport with catalytic converters.

Reassess the use of official vehicles and where appropriate find other means of transport.

Consider providing company bicycles.

Housing

Improve security on local government property with better lighting and security locks.

Ensure that all new building developments are well designed, improve the area and are sympathetic to surrounding buildings.

Provide design guides for local businesses and home owners on shop fronts, dry maintenance, roofing, windows, energy conservation, and so on.

Create a scientific information team to monitor and assess pollution-related issues within the area.

Provide testing facilities for the chemical analysis of air, water, soil, and food.

Scientific advice

Replace all CFC or halon products in use and provide disposal facilities for the public.

Monitor the safety of ground water and take appropriate steps to prosecute local businesses and water authorities that fall foul of regulations on water quality.

Arrange facilities to warn local residents of levels of radiation.

Arrange for swimming pools to phase out the use of chlorine as a disinfectant and opt for ozone (O_3) as an alternative.

Restrict the use of pesticides wherever possible.

Provide strict safety guidelines for workers using pesticides.

Introduce pesticide-free zones in public recreation areas.

Introduce codes for local building sites, including agreements on noise levels, pollution, and dust control.

Oppose new incineration plants. Work to remove chlorine sources from hospital incinerators and crematoria. Stop municipal incineration, and call for the closure of toxic-waste incinerators.

Offer dog-training classes.

Ensure that there are sufficient bylaws to regulate dog fouling.

Recycling

Ensure that the local authority has sufficient recycling facilities for all materials, including hazardous household waste.

Develop a recycling policy that reduces the amount of waste produced so that less waste has to be collected. Educate the public about reusing as much waste as possible and recycle all available materials.

Use unbleached, recycled paper for internal and external communications.

Whenever possible, encourage the use of recycling facilities at local supermarket parking lots.

Encourage local stores to promote minimal, recyclable packaging.

Provide a wood-salvaging center to encourage the public to bring, buy, and reuse timber.

Ensure that as much timber as possible is saved for recycling during rehabilitation or demolition of council-owned property.

Ban the use of tropical hardwoods obtained from nonsustainable sources.

Energy

Offer the public general, technical, and scientific information and support on energy conservation measures.

Provide maximum energy efficiency in all authority buildings.

Investigate combined heat-and-power programs for local council buildings.

Carry out energy audits on estates, educational facilities, and managed properties.

Insulate offices and managed property.

Use passive solar energy conservation when designing or planning new buildings.

Health and animal welfare

Offer organic vegetables and vegetarian food in council canteens and encourage staff to eat a healthier diet.

Adopt a fox code; foxes are not dangerous in any area, even urban foxes. Stop the persecution, hunting, and gassing of this highly intelligent creature.

Always use non-animal-tested products in preference to those tested on animals.

Discourage local shops from selling furs, ivory and other products from endangered species.

Build a center for ecological studies as part of council educational and recreational facilities.

Undertake not to purchase known cancer-causing substances.

Create a Nuclear Free Zone.

National Government

Many of the things that have to be done to safeguard our environment cannot be done without commitment from government officials. They must be lobbied, written to, and visited so that they take the environment as a priority locally, nationally, and internationally. Below is a list of just some of the important policies which should be carried out by environmentally conscious politicians. Ask your politician what he or she is doing.

Defense

Our governments spend millions on defending us, with increasing amounts being spent on high-technology warfare at the cost of some of our civil-defense plans. In an ecological crisis much of the expertise would be wasted unless we begin now to reallocate tasks and money into protection of our environment rather than aggression against other countries. Here is a tiny sample of some of the things governments can do to reallocate resources to save the planet.

Reequip NATO and Warsaw Pact engineering units for disaster relief and protection from global warming emergencies.

Convert nuclear submarines and high-flying military aircraft for long-term oceanic and atmospheric monitoring.

Change the use of our satellite monitoring system to target key areas suffering from environmental degradation problems, such as deforestation.

Pollution

Pollution from the land, sea, and air is poisoning our food, our water, and our bodies. The air we breathe and the water we drink is central to our survival, and it should be any government's top priority to ensure that it is safe and clean. The pollution that we see around us is having an even more serious effect on wildlife and on our flora and fauna. Thousands of dead seals have been washed ashore in the last few years because of a mystery virus. The pollution contamination has reached sea mammals, the top of the aquatic food chain; and we are next. The measures listed below are just some of the things that governments can do to ensure that we have a future.

Implement all laws concerning pollution.

Set up collection points for hazardous waste in every town.

Establish strict safety procedures for the handling of chemicals.

Ban any form of industrial dumping in seas, rivers, and oceans, including dumping from the shore.

Halt any further expansion of the incineration program and phase out those plants already operating.

Restrict the production of packaging that is not made with environmentally sound and recyclable material.

Clean up beaches and extend the January 1989 International Law, on the dumping of plastics at sea, to cover dumping of plastics from coastal areas, including from sewers.

Ban the import of chlorine-bleached paper and pulp derivatives like rayon.

Prosecute and fine heavily companies that pollute the environment.

Ban all CFC and halon products and provide more finance for research into alternatives.

Enforce international agreements protecting small cetaceans from pollution, overfishing, habitat destruction and tuna fishing.

Require commercial and industrial producers of waste to be licensed.

Set up enforcement teams to stop the illegal dumping of toxic waste at sea and bring those companies and individuals involved to justice.

Stop the dumping of untreated sewage into the oceans and seas and research new ways of using sewage as industrial manure.

Monitor and prosecute companies that risk the lives of local inhabitants through pollution and industrial waste from factories.

Energy

The greenhouse effect can be halted with a sensible energy policy. A government's policy should be concentrated on using as little energy as possible for the maximum return, instead of the current thinking which simply allows energy to be used with no regard for environmental concerns. Energy conservation is the first step and money and resources have to be put into researching renewable sources of energy for the future. The following recommendations represent some of the more important demands from environmentalists.

Decommission existing nuclear power stations.

Limit radioactive and polluting emissions from nuclear power stations.

End radioactive discharges into the Irish Sea.

Agree to on-site storage of nuclear waste.

Reallocate staff from nuclear power stations to research renewable energy, waste disposal, nuclear waste care, and management and environmental pollution.

Increase grants to small businesses and individuals for energy conservation measures.

Where necessary, subsidize, lend, or give grants to people receiving public welfare so that they can buy energy-efficient equipment, such as low-energy light bulbs, efficient cookers, fridges, and so on.

Enforce a national Energy Efficiency Act requiring companies and contractors to produce goods and build houses with the maximum efficiency.

Global concern

Global cooperation is crucial for our survival. Now more than ever we have realized that environmental pollution recognizes no political boundaries. Crucial to the fight to save our planet is the cooperation and help that we give to the developing countries. For so long we have used their people and land as a cheap source of labor and food that we have helped to destroy millions of acres of once productive land. Our help to them should be empowering, not tying them to a series of stringent, Western policy prescriptions, including mounting debt. Nor can we continue to use the developing countries as a toxic waste can for our industries, just because the waste is too dangerous to be dealt with on our own territory.

Develop a national least-cost plan for fuel efficiency.

Triple the research and development budgets for fuel efficiency.

Ban the cross-border shipment of any toxic or hazardous waste.

Offer scientific, financial, and political support to developing countries who wish to ban the cross-border shipment of any toxic or hazardous waste.

Only support development projects that are environmentally sustainable and take account of the needs of women.

Legislate to ensure that hardwood imports from the tropics for industry come only from sustainable sources.

Press for relief for the poorest countries on debts owed to the World Bank and International Monetary Fund and for some debts to be canceled completely.

Increase the amount of government money spent on aid to developing countries to bring us in line with Europe and the United Nations. The target is at least 0.7 percent of the Gross National Product.

Sign an international treaty to recognize Antarctica as an independent continent and help to protect it as a national park.

Use all space program facilities to research and devise new means to protect the atmosphere of the earth, especially the ozone layer.

Ban all NASA, Ariane, and Soyuz rockets, shuttles, and boosters which use freon gases and eject chlorine containing compounds that destroy the ozone layer.

Transport

The bias in government transport policy in favor of the car must be ended. The car has become one of the world's largest polluters, producing the chemicals that cause the greenhouse effect, acid rain and damage to human beings through lead and other pollutants. An understanding of the harm caused by cars must inform any new transport policy and a new approach would have to recognize that new roads, highways and cars will never solve the problems of transporting people from home to work and back again. The recommendations below are a selection of those that could start to see a regeneration of services, now almost in crisis because of current policy.

Increase corporate automobile fuel efficiency standards.

Reallocate road expenditure to provide better facilities for pedestrians and cyclists.

Phase out all leaded gas.

Research the impact of, and produce recommendations and legislation to cover, the environmental effect of diesel oils in transport.

Halt all further motorway and road plans and agree to specific environmental criteria.

Phase out the manufacture and sale of cars without catalytic converters.

Remove any further subsidies on company cars.

Lower speed limits to encourage road safety and efficient use of gas resources.

Subsidize rail and bus networks and encourage the public to use them as widely as possible.

Investigate other forms of transportation including river systems, light railways, and streetcars.

Agriculture and countryside

Farming is suffering, small farmers continue to go out of business, and the condition of our soil is becoming progressively poor due to intensive methods which are stripping the soil of its nutrients. There are many things that farmers, ecologists and even international leaders have suggested should be done to save and protect the future of our countryside. A holistic approach is now long overdue and many fear that unless intense pressure is brought to bear on Congress there may be no time left for debate.

Improve the quality of drinking water, including, if necessary, bringing the water authorities back into public ownership.

Restrict the use of nitrogen fertilizers, and observe the World Health Organization's recommended limit of 50 milligrams of nitrate in each 1.7 pints (liter) of drinking water.

Greatly increase research into nonchemical pest control.

Offer subsidies to farmers to encourage them to switch to organic farming.

Ban battery barn, semi-intensive, litter, and perchery methods of keeping chickens.

Ban growth hormones and stimulants in animals for the food industry.

End subsidies to farmers to graze protected moorlands or wetlands.

Ban the ritual slaughter of any animal.

Ban the import of inhumanely reared veal.

Encourage mixed farming.

Halt all drainage grants which destroy valuable wetlands, and protect those wetland areas remaining.

Abolish export subsidies so that excess production can be sold on the world market without subsidy.

Label all food products clearly with specific percentages of ingredients and details of the pesticides used in their production.

Label all consumer goods such as cleaning materials with their contents.

Develop agricultural policies that give priority to the needs of developing countries before ours, including halting the practice of dumping foodstuffs on the market that destroy the market for those countries.

Provide funds for and carry out proper protection of all sites of special scientific interest and environmentally sensitive areas.

Ban all forms of destructive tax-break forestry.

Legislate for planning controls for farming and forestry operations, where they affect hedges, trees, and woodlands, and the conversion of land to other uses.

Ban the use of chemicals on land that has never been treated with pesticides or herbicides, such as ancient woodlands and meadows.

Protect and increase the numbers of national nature reserves and national parks.

Grants and payments for the planting of broad-leafed trees should be increased and planting of nonnative species should be phased out.

The results of pesticide toxicity tests carried out by industry should be published and made freely available to the public.

Animal protection

Animals have been cruelly treated by big business and the argument that they do not suffer or feel pain is no longer believed. We must find a way of supporting our species on this planet without the continuous torture and inhumane practices that have been the norm in the last fifty years. Governments can provide the means for legislation to put a stop to some of the worst practices. These are just some of the protection measures that governments can take.

Ban the LD50 animal test which determines the single lethal dose of a chemical on animals.

Replace animal experiments by devising and funding humane research programs using a variety of alternative methods.

Ban bow-hunting.

Phase out the hunting and trapping of all animals.

Ensure fair representation on all state game boards from non-consumptive wildlife enthusiasts.

Provide extra funding for charities and research bodies to find alternatives to using animals in cancer research.

Require companies to stop the use of animal-testing for food, toiletry and other products.

Provide legislation to enable monitoring and protection of animals such as foxes in urban areas and amphibians, birds, and butterflies in green areas around the country.

Policy

Government policy needs to change fast. Representatives who put profit before survival need to reexamine their priorities. But policy changes that make any difference have to come from the top and the only people who can change the ideas of those at the top is the public. Letters and lobbying on specific issues raise concern but fundamental changes must be linked to a change of priorities from the public. Let politicians know how you think the planet should be saved.

Provide a requirement for environmental-impact assessments for each new project, building or decision by national government.

Incorporate environmental matters into the policy of every government department.

Provide resources and new funding for economic developments that make good use of local resources and provide for local needs.

Increase the numbers of jobs in conservation work and in scientific research on ecology and environmentalism.

Fund the development and restoration of buildings for new housing, community buildings and other uses.

Ensure that government tenders list environmental criteria as a concern. They should include assessments of how new projects may affect the environment.

Encourage less consumption.

Civil Liberties

Invest and shop for the advancement of women

Women make up 52 percent of the human population of the world. They do most of the world's work, but they own less than 1 percent of the world's land. The lives of women in developing countries have become much harder over the last twenty years, for a variety of reasons. In most countries women are not seen as men's equals. Yet at the same time they are expected to produce and care for the next generation. Women can contribute greatly to saving the planet and should be encouraged wherever possible. Invest in women-owned business and support women's cooperatives, positive schemes to aid equal opportunities and the Women's Environmental Network.

Stop corporations exploiting women in developing countries

Bottle-feeding is a major cause of child diarrhea and 200,000 deaths are recorded annually in Pakistan alone. Producers of breast-milk substitutes have been issued with an international code for marketing and selling their products in developing countries. But reports from the Philippines, Pakistan, and Malaysia tell of violations at every possible level. Malaysian groups report wholesale blackmail by companies that give the formula free in maternity hospitals, telling women that because Westerners drink it, it is the best food for their baby. This costs

269

mothers much of their badly needed income when a free and safe alternative, breast-milk, is available.

End double standards

One corporation sells the pesticide dieldrin to the Philippines for use on exotic fruits which are then exported. Workers are poorly advised about the health risks from using such chemicals thought too dangerous for our consumption. No food, chemical, technology, pesticide, or experiment that has been banned in Western countries should be dumped on the developing world. Our high standards should be applied globally. Corporations should have a duty to be responsible to the world, not just to their shareholders.

Help the homeless and hungry

Homelessness is not just a feature of the starving in Africa. Homeless families end up in cheap hotels without cooking facilities or privacy. Their environment and their bad diet mean that young children and babies especially face a serious health risk.

Support tribal peoples

Tribal people living in rain forest areas across the world are in danger. They have lived in and looked after the forest for centuries, living in balance and harmony. Today our form of "development" is destroying them. Whole communities are being wiped out by Western diseases such as measles and chicken pox, illnesses unheard of before we came along. Children are dying of lung diseases because of the dust caused by building new roads through the forest. These people know the plants and fruits of the forest intimately—that is why they have survived for so long. They have given us many of our new medicines and foods. We have no right to impose our life-style and diseases on these people. Support groups that work to save the tribal people's rights and dignity.

Campaign for disabled access

Our transportation system is dominated by roads that are built for able-bodied car drivers. Pedestrians and cyclists take second place. The section of our society that really suffers from bad access, however, is disabled people. Potholes, lack of safe places to cross, high curbs and tiny pavements are all hazards. But think about trying to get on a bus or subway with a wheelchair. Virtually impossible as well as dangerous. Have we got the right to confine people to their homes because we will not accommodate them on the streets or on public transportation?

Reclaim your right to walk

You have as much right to be walking on the road as they have to drive, and you do not cause pollution, accidents, and hazards, or waste energy. "Traffic-calming" schemes, car-free areas, and pedestrian precincts all help in urban areas.

Stand up for children's rights

Some 75 million children worldwide under the age of fifteen are forced to work for a living. They have to help to pay off their family debts, because of the costs of modernization in agriculture, the destruction of local habitats and forests, including rain forests, and the increasing burden of debt placed on farmers by the world's large banks and financial institutions. An international code of human rights includes rights for children. This should be supported around the world.

Take direct action

Direct action is not confined solely to Greenpeace activists in boats who throw themselves in front of nuclear warships. It is something we can all do, and it is easier than you think. Why not begin by telling people in the street not to drop litter if you see them? Or by sending back chemical products to the manufacturers asking them to dispose of them carefully? Or by handing out leaflets outside your hardwood store, which sells wood from rain forests? Or even by deciding to stand up yourself for the things you believe in, by boycotting products from companies that could destroy your environment? Stand up for yourself and the planet!

Fight for compensation for victims of environmental disasters

The Bhopal disaster in 1984 affected over 600,000 people, most of whom have still not received help financially, or even medically. Only 120,000 claimants have been assessed for government help so far. The company concerned, Union Carbide, has offered a final settlement of $470 million, but money is not the only problem. It will not bring back the 3,150 who died. Medical care is needed by 500,000 victims. The leak of toxic methyl isocyanate gas raises serious questions about the feasibility of companies producing such dangerous poisons so close to where people live. Highly dangerous isocyanates are used in the manufacture of resins, specialist paints and sealants. If we cut down on our use of these chemicals we can cut down on the chances of another accident.

Invest and shop for the right to know

We all need information and facts to form opinions about the environment, development, politics, and our lives in general. Companies are usually not required by law to offer us information and facts about their operations and yet we give them billions of our money. Ask companies to list ingredients, give accounts, and research results more freely so that we can make up our own minds. Support those companies that give information. Your investment in their future could stop other companies that are less open.

Don't buy cash crops

Cash crops are those commodities, from developing countries, such as cocoa, coffee, tea, and peanuts. We have engineered a system of production which means that developing countries stop growing food for themselves and start growing exotic or specialist foods for us, the richer nations. This uses up land, pesticides and severely imbalances the local ecosystem. During the worst years of the Ethiopian famine, when pictures of dead children were broadcast all over the world, the Ethiopian people had a bumper crop of peanuts but could not eat a single one. They all had to be shipped to the north to pay off crippling debts imposed by us in the first place. Only world condemnation of this sort of atrocity will stop the vicious circle of poverty for the majority of the planet's population, and only a dramatic curtailing of our desire for too great a variety of specialist foods will make a difference.

Give equal consideration to animals

Pain is pain. Nonhuman animals experience pain. As human animals we consider ourselves superior to nonhumans and inflict pain of all kinds on animals to maintain our life-styles (as in medical research and cosmetic testing), or even for pure pleasure (as in hunting and animal sports). Nonhuman animals cannot speak our language, but consideration of feelings must be the basis on which we begin to respect the nonhuman animal world. Animals feel pain and suffer if hit, drugged, deprived of food, light or water, and it is time we woke up to that. Our species superiority is not a justification. We would not consider mistreating a six-month-old baby just because it could not speak, vote, fight back, or intellectualize, so should we really do it to an animal?

Support disarmament

Trillions of dollars are spent worldwide on weapons of war that we hope we will never use, while much-needed cash that could be used to feed the hungry and protect the environment is missing. Global concern is now shifting from arms to the environment, but as yet the resources being spent on the former are not being reallocated to save the latter. Only a massive change of heart by governments across the world will stop us wasting precious resources on arms to kill. Redirect military funds towards sustainable development.

Learn the law

There are many laws affecting the environment, some of them international. The International Maritime Organization, for example, has jurisdiction on the sea and oceans, while the United Nations can issue rulings on the rights of people. Governments legislate for companies operating within their own boundaries, but they often cannot control their activities outside those boundaries. Learn the laws affecting environmental and human degradation. Without knowing what they are you will not be able to exert any influence.

Campaign for the right of common access

We need the right to walk freely in our countryside. Free access does not mean that we have the right to damage crops or walls, buildings or fences. Ramblers and walkers should avoid walking over gardens and areas where crops grow, but farmers in turn should not plow up paths for crops and should avoid the practice of putting barbed wire up to prevent access. The right of common access for those on foot should be a central part of our future countryside policy.

Commit yourself

Commit yourself to an ecological life-style: recycle waste, eat good food, repair and reuse, buy additive-free and chemical-free, think small, walk and cycle, be aware of local initiatives, avoid exploitation and use your power as a consumer whenever possible.

Talk about it

If you have read through 1,000 ways to save the planet, you will know that you can make a difference and that you are part of the problem as well as part of the solution. Others need to know this too. Talk to them, read more about the subject, do something positive. Join the millions who are organizing to save our planet.

Useful Addresses

Many nonprofit groups work on a shoestring. If you're writing to them, please enclose a stamped addressed envelope and give them a clear idea of what you want.

Alliance to Save Energy
1725 K Street, NW
Suite 914
Washington, DC 20036
(202) 857-0666

American Association of
Zoological Parks and Aquariums
Oglebay Park
Wheeling, WV 26003
(304) 242-2160

American Cetacean Society
P.O. Box 2639
San Pedro, CA 90731-0943
(213) 548–6279

American Council for an Energy Efficient Economy
1001 Connecticut Avenue, NW, #535
Washington, DC 20036
(202) 429-8873

American Wilderness Alliance
1920 N Street, NW, Suite 400
Washington, DC 20036
(202) 659-5170

American Forestry Association
P.O. Box 2000
Washington, DC 20013
(202) 667-3300

American Littoral Society
Shady Hook
Highlands, NJ 07732
(201) 291-0055

American Paper Institute
260 Madison Avenue
New York, NY 10016
(212) 340-0600

American Rivers
801 Pennsylvania Avenue, SE, Suite 303
Washington, DC 20003
(202) 547-6900

American Solar Energy Society
2400 Central Avenue, B-1
Boulder, CO 80301
(303) 443-3130

American Wilderness Alliance
7600 East Arapahoe
Suite 114
Englewood, CO 80112
(303) 771-0380

Americans for the Environment
1400 16th Street, NW
Washington, DC 20036
(202) 797-6665

Americans for Safe Food
1501 16th Street, NW
Washington, DC 20036
(202) 332-9110

Appalachian Trail Conference
P.O. Box 807
Washington & Jackson Streets
Harpers Ferry, WV 25425
(304) 535-6331

Arctic Institute of North America
University of Calgary
2500 University Drive
Calgary, Alberta, Canada
T2N 1N4
(403) 220-7515

Atlantic Center for the Environment
39 South Main St.
Ipswich, MA 01938
(508) 356-0038

Bat Conservation International
P.O. Box 162603
Austin TX 78746
(512) 327-9721

Bio-Integral Resource Center (BIRC)
P.O. Box 7414
Berkeley, CA 94707
(415) 524-2567

Caribbean Conservation Corporation
P.O. Box 2866
Gainesville, FL 32602
(904) 373-6441

Center for Environmental Education
1725 DeSales Street, NW
Suite 500
Washington, DC 20036
(202) 429-5609

Center for Plant Conservation Inc.
125 Arborway
Jamaica Plain, MA 02130
(617) 524-6988

Center for Rural Affairs
P.O. Box 405
Walthill, NE 68067
(402) 846-5428

Center for Science in the Public Interest
1501 16th Street, NW
Washington, DC 20036
(202) 332-9110

Citizens Acid Rain Monitoring Network
c/o National Audubon Society
950 Third Avenue
New York, NY 10022
(212) 832-3200

Citizens Clearinghouse for Hazardous Wastes
P.O. Box 926
Arlington, VA 22216
(703) 276-7070

Clean Water Action Project
733 15th Street, NW
Suite 1110
Washington, DC 20005
(202) 547-1196

Climate Change Activist Network
c/o National Audubon Society
950 Third Avenue
New York, NY 10022
(212) 832-3200

Coalition for Scenic Beauty
216 7th Street, SE
Washington, DC 20003
(202) 546-1100

Concern, Inc.
1794 Columbia Road, NW
Washington, DC 20009
(202) 328-8160

Conservation Law Foundation
3 Joy Street
Boston, MA 02108-1497
(617) 742-2540

Council on Economic Priorities
30 Irving Place
New York, NY 10003
800-U-CAN-HELP

The Cousteau Society
930 W. 21st St.
Norfolk, VA 23517
(804) 627-1144

Defenders of Wildlife
1244 19th Street, NW
Washington, DC 20036
(202) 659-9510

Ducks Unlimited, Inc.
One Waterfowl Way
Long Grove, IL 60047

Earth First!
P.O. Box 5871
Tucson, AZ 85703
(602) 622-1371

Earth Island Institute
300 Broadway
Suite 28
San Francisco, CA 94133
(415) 788-3666

Earthworm, Inc.
186 South Street
Boston, MA 02111
(617) 426-7344

Elsa Wild Animal Appeal
Elsa Clubs of America
P.O. Box 4572
N. Hollywood, CA 90607
(818) 761-8387

Environmental Action
1525 New Hampshire Avenue, NW
Washington, DC 20036
(202) 745-4870

Environmental Action Coalition
625 Broadway
New York, NY 10012
(212) 677-1601

Environmental Defense Fund
257 Park Avenue South
New York, NY 10010
(212) 505-2100

Environmental Law Institute
1616 P Street, NW
Suite 200
Washington, DC 20036
(202) 328-5150

Fish and Wildlife Reference Service
5430 Grosvenor Lane
Suite 110
Bethesda, MD 20814
(301) 492-6403   (800) 582-3421

The Forest Trust
P.O. Box 9238
Santa Fe, NM 87504
(505) 983-8992

Freshwater Foundation
2500 Shadywood Road
Box 90
Navarre, MN 55392
(612) 471-8407

Friends of the Earth
Environmental Policy Institute
Oceanic Society
218 D Street, SE
Washington, DC 20003
(202) 544-2600

Friends of the River, Inc.
Bldg. C
Fort Mason Center
San Francisco, CA 94123
(415) 771-0400

Fund for Animals, Inc.
850 Sligo Avenue
Suite LL2
Silver Spring, MD 20910

Global Tomorrow Coalition
1325 G Street, NW, Suite 915
Washington, DC 20005-3104
(202) 628-4016

Grand Canyon Trust
1400 16th Street, NW, Suite 300
Washington, DC 20036
(202) 797-5429

Great Bear Foundation
P.O. Box 2699
Missoula, MT 59806
(406) 721-3009

Greater Yellowstone Commission
P.O. Box 1874
Bozeman, MT 59715
(406) 586-1593

Green Mountain Club
P.O. Box 889
Montpelier, VT 05602
(802) 223-3463

Greenhouse Crisis Foundation
1130 17th Street, NW, Suite 630
Washington, DC 20036
(202) 466-2823

Greenpeace, USA
1436 U Street, NW
Washington, DC 20009
(202) 462-1177

Household Hazardous Waste Project
901 S. National Box 87
Springfield, MO 65804
(417) 836-5777

INFORM
381 Park Ave. South
New York, NY 10016
(212) 689-4040

Institute for Alternative Agriculture
9200 Edmonston Road, Suite 117
Greenbelt, MD 20770
(301) 441-8777

Institute for Local Self Reliance
2425 18th Street, NW
Washington, DC 20009
(202) 232-4108

Institute for Transportation and Development Policy
P.O. Box 56538
Washington, DC 20011
(301) 589-1810

International Institute for Energy Conservation
420 C Street NE
Washington, DC 20002
(202) 546-3388

International Oceanographic Foundation
3979 Rickenbacker Causeway
Virginia Key
Miami, FL 33149
(305) 361-5786

International Osprey Foundation
P.O. Box 250
Sanibel, FL 33957
(813) 472-5218

Izaak Walton League of America
1401 Wilson Blvd.
Level B
Arlington, VA 22209
(703) 528-1818

Keep America Beautiful, Inc.
9 West Broad Street
Stamford, CT 06902
(203) 323-8987

The Land Institute
2440 East Water Well Road
Salinas, KS 67401
(913) 823-5376

Land Trust Exchange
1017 Duke Street
Alexandria, VA 22314
(703) 683-7778

League of Conservation Voters
2000 L Street, NW
Suite 804
Washington, DC 20036
(202) 785-8683

Manomet Bird Observatory
P.O. Box 936
Manomet, MA 02345
(508) 224-6521

National Arbor Day Foundation
100 Arbor Avenue
Nebraska City, NE 68410
(402) 474-5655

National Association of Diaper Services
2017 Walnut Street
Philadelphia, PA 19103
(215) 569-3650

National Audubon Society
950 Third Avenue
New York, NY 10022
(212) 832-3200

National Coalition Against the Misuse of Pesticides (NCAMP)
530 7th Street, SE
Washington, DC 20003
(202) 543-5450

National Coalition for Marine Conservation
P.O. Box 23298
Savannah, GA 31403
(912) 234-8062

National Cooperative Business Institute
1401 New York Avenue, NW
Suite 1100
Washington, DC 20005
(202) 638-6222

National Institute for Urban Wildlife
10921 Trotting Ridge Way
Columbia, MD 21044-2831
(301) 596-3311

National Parks and Conservation Association
1015 31st Street, NW
Washington, DC 20007
(202) 944-8530

National Wildflower Research Center
2600 FM 973 North
Austin, TX 78725
(512) 929-3600

National Wildlife Federation
1412 16th Street, NW
Washington DC 20036-2266
(202) 797-6800

National Wild Turkey Feder-
ation
Wild Turkey Bldg.
P.O. Box 530
Edgefield, SC 29824
(803) 637-3106

Natural Resources Defense
Council
40 West 20th St.
New York, NY 10011
(212) 727-2700

Nevada State Commission for
the Preservation of Wild
Horses
Stuart Facility
Capitol Complex
Carson City, NV 89710
(702) 885-5589

New York Rainforest Alli-
ance
295 Madison Avenue
Suite 1804
New York, NY 10017
(212) 644-8985

North American Lake Man-
agement
P.O. Box 217
Merrifield, VA 22116
(202) 466-8550

North American Native
Fishes Association
123 W. Mt. Airy Ave.
Philadelphia, PA 19119
(215) 247-0384

Ocean Arks International
1 Locust Street
Falmouth, MA 02540
(508) 540-6801

Pacific Whale Foundation
101 North Kihei Road
Suite 21
Kihei, Maui, HI 96753
(800) WHALE-11

Pennsylvania Resources
Council
P.O. Box 88
Media, PA 19063-0088
(215) 565-9131

Project Lighthawk
P.O. Box 8163
Sante Fe, NM 87504-8163
(505) 982-9656

U.S. Public Interest Research
Group
215 Pennsylvania Avenue,
SE
Washington, DC 20003
(202) 546-9707

Rails to Trails Conservancy
1400 16th Street, #300
Washington, DC 20036
(202) 797-5400

Rainforest Action Network
301 Broadway
San Francisco, CA 94133
(415) 398-4404

Rene Dubos Center for the
Human Environment
100 East 85th St.
New York, NY 10028
(212) 249-7745

Renew America
1001 Connecticut Avenue,
NW, #719
Washington, DC 20036
(202) 466-6880

Rocky Mountain Institute
1739 Snowmass Creek Road
Snowmass, CO 81654-9199
(303) 927-3128

Sea Shepherd Conservation
P.O. Box 7000
S. Redondo Beach, CA
90277
(213) 373-6979

Ruffed Grouse Society
1400 Lee Drive
Coraopolis, PA 15108
(412) 262-4044

Share Our Strength (SOS)
733 15th Street, NW
Suite 700
Washington, DC 20005
(800) 222-1767

Sierra Club
730 Polk Street
San Francisco, CA 94109
(415) 776-2211

State and Territorial Air Pollution Program
Administration/
Air Pollution Control Officials (STAPPA and ALCO)
444 N. Capitol Street, NW
Suite 306
Washington, DC 20003
(202) 624-7864

Student Conservation Association
P.O. Box 550
Charlestown, NH 03603
(603) 826-5741

Survival International
2121 Decatur Place, NW
Washington, DC 20008
(202) 265-1077

The Trust for Public Land
116 New Montgomery St.
San Francisco, CA 94105
(415) 495-4014

Trustees for Alaska
725 Christensen Drive,
Suite 4
Anchorage, AK 99501

United Nations Environment
Program
2 United Nations Plaza
New York, NY 10017
(212) 963-8093

Water Information Network
P.O. Box 909
Ashland, OR 97520
(800) 533-6714

Water Pollution Control Federation
601 Wythe St.
Alexandria, VA 22314-1944
(703) 684-2400

The Whale Center
3929 Piedmont Avenue
Oakland, CA 94611
(415) 654-6621

Whooping Crane Conservation Association
3000 Meadowlark Drive
Sierra Vista, AZ 85635
(602) 458-0971

The Wilderness Society
1400 I Street, NW
Washington, DC 20005
(202) 842-3400

Worldwatch Institute
1776 Massachusetts Avenue, NW
Washington, DC 20036
(202) 452-1999

Worldwide
1250 24th Street, NW
Fourth Floor
Washington, DC 20037
(202) 331-9863

World Wildlife Fund
1250 24th Street, NW
Washington, DC 20037
(202) 293-4800

Zero Population Growth
1400 16th Street, NW
Suite 320
Washington, DC 20036
(202) 322-2200

About the Author

Bernadette Vallely is the director of the Women's Environmental Network and has been working on environmental issues for over five years. She has worked at Friends of the Earth and at The Anti-Apartheid Movement and has been involved with Greenpeace, SERA, and many other similar organizations.

In early 1988 she founded the Women's Environomental Network, which has since grown into an exciting and powerful organization. She is author of *Fundraising Is Fun* and *A Guide for Local Groups* and coauthor of *The Sanitary Protection Scandal* and *The Young Person's Guide to Saving the Planet*.

Bernadette Vallely is twenty-eight and lives with her husband, also an environmentalist, in London.